OIL SECURITY

OIL SECURITY
Retrospect and Prospect

Edward R. Fried
and
Philip H. Trezise

The Brookings Institution
Washington, D.C.

Copyright 1993

THE BROOKINGS INSTITUTION

1775 Massachusetts Avenue, N.W., Washington, D.C. 20036

Library of Congress Cataloging-in-Publication data

Fried, Edward R.
    Oil security : retrospect and prospect / Edward R. Fried and
Philip H. Trezise
        p. cm.
    Includes bibliographical references.
    ISBN 0-8157-2979-0 (pbk. : alk. paper)
    1. Petroleum products—Prices.   2. Petroleum industry and
trade—Forecasting.   I. Trezise, Philip H., 1912-   . II. Title.
HD9560.4.F75   1993
338.2'7282'0112—dc20   93-24983
                                                                 CIP

9 8 7 6 5 4 3 2 1

The paper used in this publication meets the minimum
requirements of the American National Standard for Information
Sciences—Permanence of paper for Printed Library Materials,
ANSI Z39.48-1984

Foreword

AMERICANS LEARNED OF an oil security problem in 1973 when crude oil supplies from the Middle East were abruptly and sharply reduced. Long lines at gasoline pumps became visible evidence that a problem existed, and within a few months the price of oil quadrupled. Declining national income in 1974–75, persisting inflation, and another oil-supply-induced recession in 1980 emphasized the disruptive potential of an interruption in Middle Eastern oil exports. In 1990, the prospect of a third oil shock contributed to the international resolve to reverse Iraq's invasion of Kuwait and its threat to Saudi Arabia's oil fields by military means.

Looking to the future, the authors of this book find that while the possibility of new oil shocks cannot be dismissed, the damage that might be visited on the world economy can be limited in all but a worst-case scenario at comparatively modest cost. Demand and supply fundamentals alone promise a relatively stable real price for oil lasting well into the first decade of the next century. Oil markets in the big consuming countries are far more efficient than they were in the 1970s. And emergency stocks of crude oil held by members of the Organization for Economic Cooperation and Development, if resolutely used, can provide a critical offset to any substantial reduction in oil shipments from the Middle East.

Edward R. Fried and Philip H. Trezise are senior fellows in the Brookings Foreign Policy Studies program. They are indebted to Thomas G. Burns, A. Denny Ellerman, Thomas L. McNaugher, Charles D. Masters and William B. Quandt for insightful comments on the manuscript. Theresa Walker edited the manuscript, and Adrianne Goins verified it. Ann Ziegler provided secretarial assistance.

An earlier version of the manuscript was prepared as a Tokyo Club Paper, and the Brookings Institution is grateful for the financial assistance of the Tokyo Club Foundation for Global Studies.

The views expressed in this book are those of the authors and should not be ascribed to any of the persons or the funding source acknowledged above, or to the trustees, officers, or other staff members of the Brookings Institution.

Bruce K. MacLaury
President

July 1993
Washington, D.C.

Contents

OIL SECURITY

CHAPTER ONE

A Less Troubled Future?

THIS BOOK IS a modest exercise in peering into the world's oil future, specifically to assess the oil security problem over the medium term. Frequently the term *energy security* is used to describe the problem. This can be misleading. Other energy fuels—coal, natural gas, nuclear, and renewables—constitute more than 60 percent by volume of the world's sources of commercial energy. No special concern, however, attaches to the security of their supply or to their pricing. When oil price shocks occurred in the 1970s, prices of these other fuels also rose, but only after a lag, and not by as much or for as long. Only another interruption in the supply of oil and the consequences flowing from it define the so-called energy problem.

This study therefore focuses on oil. It finds that market fundamentals in themselves promise that the proximate future should be free of the sudden and discrete price increases that have made oil security a term that encompasses large economic, political, and even military concerns for oil-importing nations. Supply and demand considerations point to stable oil prices, in the range of $20–$25 per barrel (in 1991 dollars), continuing far into the first decade of the next century. New, widely applied taxes or other measures to avoid environmental degradation, which cannot now be taken into account, would fortify this comparative optimism about the price of oil.

Political or military events in the principal Middle East exporting countries could undo this prospect. Oil supplies, as has been expensively learned, can be subject to unanticipated and sizable interruptions. In the 1970s and 1980s such interruptions proved extraordinarily costly to the world economy, so much so that in 1990 the threat of a recurrence led to an unprecedented international military intervention in Iraq.

No crystal ball is available to tell us whether or when the supply of oil might again be put at risk. What can be hazarded is that the potential exists for containing the damage from another episode. The underlying forces making for oil price stability are a favorable factor. So is the fact that domestic oil market distortions have been largely eliminated, at least in the major industrial countries. Most important, comparatively modest improvements in the defensive system designed in the 1970s by the members of the International Energy Agency (IEA) can insure against the high-cost disruptions experienced in and after the 1973–74 and 1979–80 shocks. Put another way, except in an improbable worst case the likelihood of profoundly damaging shocks to the world economy from a sudden jump in oil prices need not be large.

Dependence on oil from the Organization of Petroleum Exporting Countries (OPEC) will not diminish. Rather, OPEC, especially its Persian Gulf members, will supply an increasingly large share of the world's demand for oil between now and the year 2010. This prospect, we believe, does not conflict with a realistic hope for an extended respite from the damagingly volatile oil prices seen during the past two decades.

Nor is oil's key role in the world economy in question. Indeed, the value of oil production exceeds that of any other primary commodity. At the time of the invasion of Kuwait, when oil was $22 a barrel, the production of primary energy fuels—oil, coal, natural gas—constituted 3.3 percent of world GNP, with oil accounting for two-thirds of that total. The value of oil production was one and one-half times that of the world's production of food grains (wheat, rice, coarse grains). Oil exports were about 7 percent of world exports, ten times as great as the value of food grain exports. These relationships may change but not soon.

Oil, furthermore, is special among the principal tradeable commodities in the delayed response of demand and supply to higher prices. Given these low elasticities, price run-ups can persist for extended periods if the leading producers choose to maximize prices at the expense of volume and can effectively collude to that end. In the 1970s and again in the 1980s, greater efficiency in oil use, the substitution of other fuels for oil, and the development of new supplies combined eventually, but only after several years, to force prices down again. That is not the typical commodity case. Grain prices tripled in the crop year 1972–73 owing to a worldwide drought, then fell to predrought levels as production recovered in the next crop year. In 1973–74 prices for industrial raw materials also soared after a lengthy industrial boom. Fears that these high prices would last were quickly proved wrong as demand declined and new supplies came on stream. Quantitatively, the grain and raw material price increases had a much smaller immediate impact and the subsequent economic adjustments came much faster than was true of oil.

Oil is different, too, in that the world's resources are heavily concentrated in the Middle East. Whether the focus is on reserves, present output, exports, or potential spare capacity for emergencies, the region's oil fields have a dominant position in the world's oil economy. Worrisome political fissures in the Middle East make oil supply interruptions a continuing possibility and a persistent concern to policymakers in the United States and other major importing nations.

A glance at the recent past explains why. In 1970 when oil was cheap and its consumption growing rapidly, primary energy fuels contributed 2.0 percent of the value of world production at that time, with oil 1.3 percent of that total. In 1980, when the world price of oil peaked, primary fuels constituted 9.4 percent and oil 6.7 percent of world GNP. Such large swings in a comparatively brief period set in motion vast structural, financial, and macroeconomic changes around the world. Interruption or the threat of interruptions of Persian Gulf oil in 1973–74 and 1979–80 caused widespread disturbances—inflation, unemployment, foregone production—and recovery proved slow and painful.

These events followed two golden decades of strong economic growth worldwide — growth associated with low and declining oil prices. The oil shocks, which to many people signified a discrete break with this retrospectively untroubled past, aroused all manner of fears, justified and imaginary. The revival of raw materials Malthusianism exemplified in the Club of Rome's *Limits to Growth* was one myth.[1] A related proposition held that oil would be the forerunner of a proliferation of commodity cartels. Another flight of imagination saw Arab sheiks with their newfound wealth dominating and manipulating international financial markets. And control of the oil supply suddenly became critical in calculations of political influence and power in the world.

The reality of course was less cosmic. For the oil-importing countries disruptive, costly consequences inevitably followed huge, unanticipated percentage increases in the price of a widely used commodity. For the exporters the resultant windfall gains, to be squandered or used prudently, did not confer the far-reaching leverage that the alarmists feared. For both, the oil shocks taught commonsense lessons.

For one thing, oil security for any country does not mean achieving greater self-sufficiency in oil or, even as the Nixon administration promised, reaching a state of "energy independence." Rather, security lies in the avoidance or mitigation of a sudden, substantial, and potentially prolonged rise in the world price of oil. Such a rise would almost certainly impose costs that are economywide, protracted, and high.

Public policies can moderate these costs. But, as the United States in particular learned, that effect is not true of policies intended to control prices by legislative or administrative fiat. Nor, as France and Japan discovered, can a market economy insulate itself from a global price increase by making local investment deals or long-term supply contracts with ex-

---

1. Donella H. Meadows and others, *The Limits to Growth: A Report for the Club of Rome's Project on the Predicament of Mankind* (Universe Books, 1972).

porting countries. Where arrangements of this sort were made, the importing nations found themselves buying oil at the same price others paid, or even at a premium. In these instances the importing countries failed to help themselves and in some measure worsened the position of other oil importers.

Much the same comment applies to the recurring notion that the United States somehow can gain oil security by obtaining preferred access to the oil exports of the Western Hemisphere. A like indicator of the misplaced preoccupation with bilateral oil relations is found whenever the threat of an interruption of oil supplies from the Middle East appears; the predictable U.S. media reaction is to ask how much oil is at risk for the United States. This kind of thinking is symptomatic of the failure to recognize that oil security for any country is determined only by developments in the world oil market and specifically by the price reactions that occur there.

Even a favorable oil resource position will not shelter a country from an oil shock. The citizens of Norway and the United Kingdom, countries that are net oil exporters, avoided terms-of-trade losses from the oil shocks but were adversely affected by increased domestic inflation and economic slowdown abroad in the same way as were people in Japan, a country that imports almost all of its oil.

For at least some of the exporting countries the 1970s and 1980s made the central point that high prices create pressures for economies in oil use that can have a lasting effect on demand. Acceptance of this elementary lesson, and credible policies among the major importing nations for dealing with contingencies, can help greatly to reduce damage from a new oil supply interruption.

In this book we examine trends in the world oil market. Demand in the OECD countries was heavily affected by the oil-use efficiencies dictated by past oil shocks. Will slow growth of demand persist in an era of relatively stable prices? Importing developing countries and especially Eastern Europe responded rather sluggishly to world prices and helped to sustain demand for OPEC oil. Will the non-OECD countries pursue oil and energy efficiency with greater enthusiasm in the years ahead?

Developments in the supply side of the market will depend primarily on investment policies among the Middle East producers, Saudi Arabia foremost. Prospects for increased output and new finds elsewhere must also be considered, as must trends toward declining reserves in the United States and the North Sea.

Environmental concerns, especially about a global warming phenomenon, now command international attention. Taken at face value, countermeasures to this and other environmental threats would call for sizable reductions in the use of and demand for oil. We seek, therefore, to assess these possibilities.

Possible surprises on the supply side are to be found mainly in the politics of the Middle East. How will the Gulf War, the end of the cold war, and the ongoing peace negotiations in the Middle East affect possible threats to the supply from that region?

Another major oil shock, if it were to occur, would take place in circumstances different than the ones in the shocks of the 1970s and the minishock of the 1990s. What happened in the earlier episodes, however, would also shed light on the potential vulnerabilities of the future oil market.

CHAPTER TWO

Oil Market Trends under Surprise-Free Assumptions

THAT OIL PRICES could play havoc with the world economy would have surprised observers of the oil industry in the 1950s and 1960s. Important new discoveries in the Middle East (as well as in the North Sea, Alaska, and Nigeria later in the period) placed the industry in a position of chronic glut even as it experienced a strong increase in demand. World market prices were low and fairly stable in the 1950s and even fell by about one-third in the 1960s, from $1.63 to $1.27 a barrel (as measured by spot prices for Saudi Arabian light oil).

Cheap oil thus contributed to strong worldwide economic growth, which in turn spurred oil consumption. Furthermore, because of price and convenience in use, oil was increasingly substituted for coal. Between 1960 and 1972, world consumption of oil increased by one and one-half times, or on average by 7 percent a year. World economic growth meanwhile measured 5 percent a year. In rapid order, the world's industries, transportation, commerce, and households became increasingly dependent on the use of oil (making the subsequent adjustments away from oil more costly).

Not surprisingly, therefore, the first oil shock marked an important divide in world attitudes toward oil and energy. The fourfold increase in price between October 1973 and January 1974 came as a traumatic break with the past, raising concerns about a new era of extreme price volatility, biased upward. At the same time, the Arab oil embargo brought into question the long-term reliability of oil from the Middle East, widely seen, even then, as key to future availability. The second

7

shock, only five years later, seemed to confirm these fears. In their aftermath, world demand for oil lost its former exuberance, rising less than 1 percent a year over the period 1973–91. The effect on consumption varied markedly, however, among the major categories of consuming countries.

In the OECD countries, the oil price hikes, supplemented by tax, regulatory, and subsidy policies to conserve oil, led to a sharp decline in the use of oil per unit of output (that is, oil intensity) and even an absolute decline in total consumption of oil over eighteen years. In the developing countries, however, higher oil prices failed to prevent oil consumption from expanding almost as much as their gross domestic product (GDP) increased. Oil consumption is now double what it was in 1973, the increase more than offsetting the decline in demand in the OECD. In the former Soviet Union and Eastern Europe, much as in OPEC countries, the oil shocks for all practical purposes did not happen. Official policy kept domestic oil prices low, which permitted oil consumption to rise by one-fifth, despite slack economic growth over the period. This policy also resulted in a huge loss of the income that could have been realized had a larger share of production been diverted to the much higher-priced international market.

Consequently, today's structure of the world's demand for oil is much different than it was in 1973 (table 2-1). The OECD share has declined sharply from 71 percent to 56 percent, raising the question of whether oil intensity will continue to fall or even be reversed if oil prices remain comparatively moderate and stable over the indefinite future.

Even so, the key variables are elsewhere. In any likely circumstances, the developing countries will continue to be the dynamic element on the demand side of the world oil market. Their share of world oil consumption, which has risen dramatically since 1973, is now approaching 30 percent. Continued high economic growth is expected for this group of countries, which will lead to increased oil consumption. To what extent will they achieve efficiency gains that will slow down the growth in their oil consumption as their incomes rise?

Table 2-1. *World Oil Demand, 1973, 1991*

Region	Oil consumption (million barrels per day) 1973	1991	Share of total (percent) 1973	1991
OECD				
France	2.6	2.0	4.6	3.0
Germany	3.4ᵃ	2.8	6.0	4.3
Japan	5.5	5.3	9.6	7.9
United Kingdom	2.3	1.8	4.0	2.6
United States	16.9	16.2	29.6	24.6
Other	10.0	9.1	17.5	13.7
Total	40.7	37.2	71.3	56.1
Non-OECD Europe				
Former Soviet Union	6.6	8.4	11.6	13.4
Other	1.2	1.4	2.1	2.1
Total	7.8	9.8	13.7	15.5
Developing countries				
China	1.1	2.4	1.9	3.7
OPEC	1.5	4.6	2.6	7.0
Other	6.0	11.5	10.5	17.6
Total	8.6	18.5	15.0	28.3
World total	57.1	65.5	100.0	100.0

Sources: *BP Statistical Review of World Energy, June 1992* (British Petroleum Company, 1992), p. 8; *BP Statistical Review of World Energy, June 1985* (London: British Petroleum Company, June 1985), p. 8; *OECD Economic Outlook*, no. 51 (Paris, June 1992), p. 152, table 78; and *World Oil Trends*, 1988–89 edition (Houston and Cambridge: Arthur Andersen & Co. and Cambridge Energy Research Associates, 1988), p. 24.
a. Includes East Germany.

Similarly, the course of oil consumption in the former Soviet Union is highly uncertain. Oil intensity there is extraordinarily high while the economic outlook is bleak. How much could economic reforms, when they are finally in place, increase the efficiency of oil use and encourage the substitution of other primary fuels for oil as the economy begins to recover?

The oil shocks were also associated with important changes on the supply side of the market as OPEC member governments took over from international companies the responsibility for determining their production levels. When OPEC also attempted to set price, it became the residual supplier to the world market. Other countries were free to produce at capacity while OPEC members, acting together as a cartel or individually, had to restrict output to support their price targets.

At the time of the oil shocks, when actual or threatened

interruptions in supply brought about a tight balance between the call on OPEC's oil and its sustainable capacity, coordinated restrictions on production were unnecessary; producers became price takers and accepted the windfalls resulting from market panic. As demand declined in the aftermath of the price jumps, OPEC members sought to contain the erosion in prices through restraining production, first through loose consultations and eventually by collectively allocating quotas for each member country. As excess capacity grew, prorationing succumbed to cheating on the quotas by most members, Saudi Arabia being the notable exception. Weakening markets called for still more attempts to stabilize the market by establishing a new set of production quotas.

Prices declined haltingly and gradually until 1986, when the call on OPEC as residual supplier fell to little more than half of its sustainable capacity. Saudi Arabia, finding itself at the low point producing at only one-fifth of its capacity and acting involuntarily as balance wheel to the system, finally gave up the struggle. It began to cut prices and compete aggressively to regain its historical share of the market, in effect flooding the market to shock other OPEC members into a new prorationing arrangement. The subsequent collapse in prices saw the entire price effect of the second oil shock and two-thirds of the first wiped out (figure 2-1). It then took time for OPEC to reestablish quotas on more realistic, although still fragile, lines and await the strengthening of demand for the recovery of prices.

While falling OECD demand took the secular buoyancy out of the market, world oil consumption except for the years 1974–75 has always been above the 1973 level. By 1991 consumption had increased 15 percent over 1973. The steady rise in non-OPEC production was the proximate cause of the increase in OPEC excess capacity and the erosion of prices that followed. A comparison of world oil production in 1973 and 1991 (table 2-2) shows the sources of this additional supply.

Oil production in the OECD countries increased moderately over the period, with higher North Sea production more than offsetting a decline in the United States. Oil fields in these

Figure 2-1. *Crude Oil Prices, 1970–91*

U.S. $ per barrel

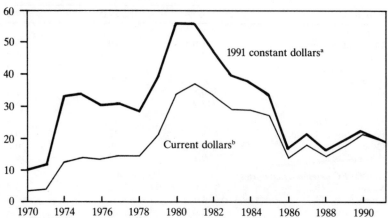

Sources: Energy Information Administration, *Monthly Energy Reivew, April 1992*, DOE/EIA-0035(92/04) (Department of Energy, April 1992), p. 105; and *Economic Report of the President Transmitted to the Congress January 1993, together with the Annual Report of the Council of Economic Advisers* (Government Printing Office, 1993), p. 352.

a. Current prices converted to 1991 constant dollars on basis of U.S. GNP deflator.

b. Average U.S. refiner acquisition cost of imported oil.

countries are mature; have they already peaked or are they soon to do so?

Developing countries, a large group that includes net importers and net exporters, have supplied the principal growth in non-OPEC oil. Their production has consistently grown year to year increasing threefold over the period as a whole. With some notable exceptions, however, reserve-to-production ratios are comparatively low, and the production base could decline by the end of the century.

In the former Soviet Union, production rose strongly during the 1970s, stayed at a high plateau in the 1980s and fell sharply beginning in 1990, probably owing to disorganization in the economy and the oil industry, shortages in equipment and supply, and poor maintenance, rather than to lack of resources.

That leaves the focus of attention on the OPEC countries, where despite huge proven reserves, capacity has not been maintained, even setting aside war damage in Iraq and Kuwait.

Table 2-2. *World Oil Production, 1973, 1985, 1991*[a]

Region	Million barrels per day		
	1973	*1985*	*1991*
OECD			
United States	11.0	10.5	9.0
Canada	2.1	1.8	2.0
United Kingdom	...	2.7	1.9
Norway	...	0.8	1.9
Other	0.8	1.2	1.2
Total	13.9	17.0	16.0
Non-OECD Europe			
Former Soviet Union	8.7	11.9	10.4
Other	0.4	0.4	0.3
Total	9.1	12.4	10.7
Non-OPEC developing countries			
Mexico	0.6	3.0	3.0
China	1.1	2.5	2.8
Other	2.6	5.4	7.2
Total	4.3	10.9	13.0
OPEC			
Saudi Arabia	7.7	3.7	8.6
Iran	5.9	2.2	3.3
Iraq	2.0	1.4	0.2
Kuwait	3.1	1.1	0.2
United Arab Emirates	1.5	1.4	2.6
Venezuela	3.5	1.7	2.6
Other	7.6	5.8	7.3
Total	31.3	17.3	24.8
Total world	58.5	57.5	64.2

Source: *BP Statistical Review of World Energy, June 1992*, p. 5; and *BP Statistical Review of World Energy, June 1984* (London: British Petroleum Company, June 1984), p. 5.

a. Includes crude oil, shale oil, oil sands, and natural gas liquids.

Will OPEC investment in new capacity take place on a large scale, or will the oil needs stemming from world economic growth be met principally through adjustments in demand compelled by higher prices?

The Demand Side: Prospective Changes in Oil Intensity

Indexes of oil intensity over time are only a rough proxy for changes in the efficiency of oil use. Changes reflect the combined effects of technological improvements (for example, increases in automobile mileage per gallon), structural differences (for example, movement toward or away from oil-intensive sectors), and fuel substitution (for example, replacement of oil by other primary fuels in power generation). The price of oil and growth in GDP underlie these changes, but as a recent study points out, autonomous factors independent of either can also affect them. Furthermore, since the replacement of the huge stock of oil-using equipment occurs only gradually, past price changes keep affecting the demand for oil for a long time, even without changes in other influences. Thus projecting the future of the demand for oil remains a hazardous exercise. No wonder past predictions have so often missed the mark.[1]

Will the Decline in OECD Oil Intensity Continue?

In 1991 OECD countries required about 40 percent less oil to produce a unit of output than in 1973, for an average gain in efficiency of 2.9 percent a year (table 2-3). Over that period total oil use in the OECD fell absolutely by 9 percent in the face of an increase of 59 percent in GDP. An oversimplified calculation makes it easier to understand the magnitude of that transformation: if the relationship between oil use and output had remained the same as in 1973, present OECD oil requirements would be about 24 mmb/d (million barrels per day) higher than they are, which would have required another OPEC to supply.

All the major OECD countries participated strongly in this transformation, although timing and degree varied among them. Most strikingly, U.S. oil intensity declined very little after the first oil shock because controls on domestic oil prices partly insulated the consumers from the jump in world prices. These controls were phased out between 1978 and 1981 so that

---

1. For a helpful description of the principal influences on oil demand and an explanation of the causes for wide discrepancies in projections, see Energy Modeling Forum, *International Oil Supplies and Demands*, Report 11, vol. 1 (Stanford University, September 1991), pp. 21–27.

Table 2-3. Trends in OECD Oil Intensity, 1973–91

Oil use per unit of real GDP

Region	Indexes, 1973 = 100				Annual rates of change (percent)			
	1978	1985	1988	1991	1973–78	1978–85	1985–91	1973–91
Total OECD	91.9	62.3	60.6	59.7	−1.7	−5.7	−0.7	−2.9
United States	98.4	67.4	66.0	63.3	−0.3	−5.6	−1.0	−2.6
Japan	72.7	51.6	51.2	48.8	−6.6	−5.0	−0.9	−4.1
Germany	87.0	61.4	58.9	54.4	−2.8	−5.1	−2.0	−3.4
France	81.9	53.2	49.5	50.5	−4.1	−6.3	−0.9	−3.9
United Kingdom	80.0	59.6	53.7	55.5	−4.6	−4.3	−1.2	−3.3

Sources: *OECD Economic Outlook*, no. 48 (Paris, December 1990), p. 35; and authors' estimates.

the second oil shock plus the effect of domestic decontrol passed through the economy. Oil intensity fell accordingly. True, for the period as a whole, oil intensity in the United States fell almost one-third less than the average for the other major countries. Structural differences can explain part of this shortfall. Differences in tax policy probably operated as well; since 1973 restrained increases in taxes on gasoline in the United States contrast with aggressive increases in Japan, Germany, France, Italy, and the United Kingdom.[2]

The decline in OECD oil intensity was close to spectacular during the 1978–85 period when the first two oil shocks exerted a cumulative effect. Similarly, during the next six years, after oil prices had fallen an average of 40 percent, the rate of decline fell almost as rapidly, nearly disappearing at the end of the period. That oil intensity continued to decline at all in the face of so large a fall in oil prices attests to the long-lasting effects of the previous price shocks.

Nonetheless, at current oil prices, the end of these gains in efficiency may be near. Might they even be reversed? Further, will the responses to any recovery of prices in the future be as large as in the past? In an econometric study of the data for the United States and Japan and a review of other studies, Dermot Gately concludes that changes in response to increases and decreases in the price of oil are not symmetrical. His data show that the demand responses to the oil price jumps of the 1970s were five times larger than the responses to the oil price cuts of the 1980s. "Thus, at most," he states, "a small fraction of the demand reductions will be reversed." His study, however, provides no clear answer to the question of whether the

2. For example, between the first quarter of 1978 and the first quarter of 1982 the federal excise tax on gasoline in the United States increased from $0.12 to $0.14 a gallon. State and local taxes rose somewhat more but even now amount to only $0.15 to $0.20 a gallon. In contrast, from 1978 to 1991 gasoline taxes in the next five largest OECD countries increased from an average of $1.00 a gallon to approximately $2.60 a gallon. Energy Information Administration, *International Energy Prices 1978–1982*, DOE/EIA-0424 (82) (Department of Energy, January 1984), pp. 38–39; and "Germany, Japan Increase Gasoline Taxes," *Oil Daily*, March 21, 1991, p. 3.

demand responses to any price recovery in the 1990s would be as large as those in the past.[3]

The contrasting behavior of oil consumption in products where fuel substitution is readily feasible and those where it is limited provides another insight. Since 1973 higher oil prices and concern about reliability of supply have practically halted the construction of new oil-fired power plants and reduced the use of existing plants; thus coal and natural gas have steadily substituted for oil in power generation. (The use of nuclear power grew autonomously, having been set in motion well before the oil shocks occurred.) In 1973 oil was the primary fuel source for 26 percent of electricity production in the OECD countries compared with about 10 percent in 1989. Use of oil in the transportation sector, however, increased by almost a quarter after 1973: a decline of 27 percent in gasoline consumption per car could not fully offset a nearly two-thirds increase in the number of cars.[4] Similarly, the use of oil as a chemical feedstock, where substitution is also difficult, grew by over 10 percent despite improved efficiencies in the manufacturing processes. It follows, therefore, that the share of oil consumption for products in which fuel substitution is limited has been growing steadily in the OECD countries, leading the OECD secretariat to suggest that "more of future reduction in oil intensity will have to come through developing and adopting more fuel-efficient designs."[5]

3. Dermot Gately, "Oil Demands in the U.S. and Japan: Why the Demand Reductions Caused by the Price Increases of the 1970's Won't Be Reversed by the Price Declines of the 1980's," Economic Research Reports, RR92-09 (New York University, C. V. Starr Center for Applied Economics, February 1992), p. 25.

4. Data on OECD oil use in electricity production and in automobiles appear in International Energy Agency, Energy Policies of IEA Countries: 1990 Review (Paris: OECD/IEA, 1991), pp. 122, 128, tables A13, A19. Data for gasoline consumption per car reflect changes in fuel efficiency and the average distance that each car is driven in a year.

5. OECD Economic Outlook, no. 48 (Paris, December 1990), p. 45. This section also points out that limited possibilities for fuel substitution in particular products does not preclude future fuel savings. Reactions of transportation product prices to changes in crude oil prices have been dampened by the high proportion of

World oil forecasts, explicitly or implicitly, unfortunately offer a range of answers to the question of what course oil intensity in the OECD will take. The International Energy Agency's (IEA) world energy outlook published in 1991 offers a plausible measure of the response of oil demand in the OECD to any given increase in GDP and in crude oil prices.[6] The IEA assumes that GDP will rise by 2.7 percent a year and that crude oil prices in real terms will rise on average by 3.5 percent a year through 2005. OECD oil intensity is then estimated to decline by 2 percent annually, or about one-third the average rate of decline in the seven years following the second oil shock. To test the sensitivity of its analysis to lower oil prices, the IEA assumes, alternately, that oil prices will remain constant at $21 a barrel in 1990 dollars between now and 2005. OECD oil demand would then grow by 2 percent a year and by 2005 would be almost 9 mmb/d higher than in the reference case. Even so, OECD oil intensity would still decline, although at a much reduced rate—by 0.7 percent a year, about the rate of decline during the past six years. Uncertainty aside, these results reemphasize the importance of the sluggish turnover of capital stock; the oil efficiency of new, technologically advanced transportation and industrial equipment over the next twenty years is estimated to continue to be higher than the average oil efficiency of the plant it replaces.

When Will the Developing Countries Show a Significant Downturn in Oil Intensity?

Throughout the period 1973–91, energy consumption in the developing countries increased more rapidly than GDP, as

---

taxes they contain. However, long-term price elasticities in the transportation sector are not markedly lower than in other sectors.

6. IEA, *Energy Policies of IEA Countries*, chap. 8. IEA estimates of oil price increases in real terms through 2005 are high in relation to the discussion of the world market outlook on pp. 38–43 in this book. They are cited here to derive oil intensity relationships under different price scenarios.

it had in the previous decade. When primary energy prices rose in the wake of the oil shocks, energy intensity showed little response.

Consumption of oil did not follow a much different course. Even after excluding the OPEC countries where domestic oil prices remained very low, oil use for the group kept pace with GDP between 1973 and 1979 and after that rose about two-thirds as fast—a remarkably restrained reaction to higher oil prices compared with the OECD experience. A World Bank in-depth study of the energy sectors of eight countries (including China and India) suggests that structural change accompanying sustained growth from a low base is perhaps the main reason. Countries covered in the study account for half the energy and one-third the oil consumed by all developing countries. With varying intent and success, these nations pursued policies that sought to restrain the use of oil and encourage indigenous alternatives.[7]

Usually, the World Bank study points out, energy and oil intensities rise as countries pass through the early stages of industrialization and urbanization and peak only as economies mature. It cites several reasons. First, an early concentration on energy-intensive heavy industries takes place. Plants tend to be antiquated and the high cost of capital discourages the introduction of the most technologically advanced equipment. As economic growth takes hold, larger, more efficient facilities and better management improve fuel efficiency, but the lag seems surprisingly long. Second, urbanization spreads rapidly and in that process traditional fuels (biomass) are replaced by kerosene and electricity, some of it oil fueled. Third, and most important, these countries are characterized by an "unstoppable need for mobility" and hence an explosive increase in cars and trucks, whose domestic manufacture initially at least produces vehicles with lower fuel efficiency than in OECD

7. Mudassar Imran and Philip Barnes, "Energy Demand in the Developing Countries: Prospects for the Future," Staff Commodity Working Paper 23 (Washington: World Bank, 1990), pp. 1–3.

countries. Interestingly, the study found car ownership to be higher among these countries than in the lower-income OECD countries when they were at similar levels of GDP per capita, probably because of the rapid decline in car prices in real terms over the postwar period. In all, by 1990 transportation accounted for 40 percent and industry for 17 percent of total oil use in the countries covered by the study.[8]

This is not to ignore policy failure as a cause for the continued rapid growth of oil use. The World Bank's *World Development Report 1992* notes that underpricing electricity in developing countries is the rule. "Prices, on average, are barely more than one-third of supply costs and are half those in industrial countries."[9] Kerosene prices are widely subsidized and transportation fuels have only recently tended to be priced on the basis of international oil prices and are only beginning to be subject to significant taxation.

What about the future? The study by Mudassar Imran and Philip Barnes applied the World Bank projection of 5.1 percent annual growth for all developing countries and of an annual real increase of 1.5 percent in oil prices to a sample group of countries. In the most likely scenario, the report also assumes a more rapid introduction of oil-efficient technology than in the past two decades and a determined move toward market-based pricing of oil products. Offsetting factors are a projected fourfold increase in automobiles and the continued substitution of commercial for traditional fuels accompanying further urbanization. The result is a doubling of oil consumption by 2010,[10] an average annual increase of 3.6 percent—or 70 percent of the projected rate of increase in GDP. Extrapolating these results to all developing countries would mean

8. Ibid., pp. 6, 70. This study found that China's chemical and metallurgy industries "are extremely fuel inefficient." Even in Brazil in the early 1980s, "the six most energy-intensive industries used on average 50 percent more energy per unit of output than in Sweden" (p. 19).

9. *World Development Report 1992: Development and the Environment* (Oxford University Press for the World Bank, 1992), p. 116.

10. Imran and Barnes, "Energy Demand," pp. 49–68. Under business-as-usual assumptions oil demand is projected to triple.

an increase by 2010 of 18 mmb/d in world oil consumption from this sector of world demand. The IEA projections, using assumptions of lower economic growth and much higher oil prices than the World Bank study, show an increase in oil consumption of developing countries of 11 mmb/d for 2005, or by extrapolation, about 14 mmb/d between now and 2010.[11]

Are these plausible projections? The relationship between increases in oil consumption and GDP growth are roughly the same as that experienced in the 1980s, which leaves too little room for improvements in efficiency. With assumptions about economic growth and oil prices given, the rate of increase of oil consumption hinges largely on the speed of technological change as more of these countries move strongly on the path of rapid industrialization. Only limited data are available to test the effect on oil consumption. In South Korea, a prime example of a large, export-driven, rapid-growth country, oil intensity between 1980 and 1990 fell 2.8 percent a year, a respectable showing even by OECD standards. Also, the *OECD Economic Outlook* estimates that oil intensities for all developing countries are 1.8 times the OECD average, but those for the high- and middle-income oil-importing developing countries are only 1.3 times the OECD average.[12] Under any circumstances, the developing countries will again be responsible for most of the growth in world oil demand through 2010, but these data suggest that the World Bank and IEA projections are much too cautious in accounting for the effect on oil consumption of the technology and pricing improvements that seem to be in the offing.

How Large Are Potential Oil Savings from Greater Efficiency of Oil Use in the Former Soviet Union?

In 1988 oil consumption in the former Soviet Union amounted to 8.8 mmb/d, somewhat more than half that of the

11. IEA, *Energy Policies of IEA Countries*, p. 64.
12. Developing countries include all non-OECD countries except the former Soviet Union and China. *OECD Economic Outlook*, no. 48, p. 42, table 19.

United States and four-fifths more than Japan. Those data alone show that oil intensity is very high, but the exact intensity is difficult to calculate because GDP figures, particularly when converted to U.S. dollars for comparability, are so uncertain. The OECD estimates, heroically, that in 1988 oil intensity was 3.1 times that in the OECD, or by implication 2.8 times that of the United States.[13] That estimate would imply that GDP per capita in 1988 was about one-sixth that in the United States, a figure that now at least seems fairly plausible; the OECD estimate of oil intensity in the former Soviet Union, therefore, may be useful as a way to think about the problem.[14]

Such profligacy in the use of oil suggests that the impact of the former Soviet Union on world oil will depend as much on reform of its domestic oil market as on rebuilding its production. There is a long way to go. Even with the price increases of May 1992, the *Economist* finds that Russia's oil is priced domestically at one-fifth the world level.[15] Moving decisively toward market-based oil prices, replacing oil with natural gas, changing the emphasis from heavy industry to the rather neglected light industry and service sectors, and infusing those sectors with new technology should bring about very large savings in oil consumption. If such changes could cut in half the current gap in oil intensity with the United States during the next two decades, oil consumption in the former

13. Ibid., p. 43, table 20.

14. The definition of "plausibility" has changed drastically in recent years. In 1981 Herbert Block, an American expert on the USSR economy, estimated that in 1979 Soviet per capita GDP was 42 percent that of the United States. See Herbert Block, *The Planetary Product in 1980: A Creative Pause?* (Department of State, 1981), pp. 30–33, app. table 1. Block's estimate was pretty much at the low end of speculation about the subject at that time. *The Economy of the USSR*, a 1990 joint report of the International Monetary Fund (IMF), the World Bank, the OECD, and the European Bank for Reconstruction and Development (EBRD), now considered outdated, ventures that Soviet per capita GNP was only $1,780 in 1989 dollars. That would mean that Soviet oil intensity was almost six times that of the United States. See *The Economy of the USSR: Summary and Recommendations* (Washington: World Bank, 1990), p. 51, app. table 3.

15. The price is at July 1992 exchange rates. "Two Steps Forward, One Step Back, One Step Sideways," *Economist*, July 4, 1992, p. 64.

Soviet Union would be 2-3 mmb/d lower than it otherwise would have been.

Meanwhile, oil consumption in the former Soviet Union is falling fast in the wake of economic disorganization and plunging economic output. PlanEcon, a research organization specializing in the former Soviet Union and Eastern Europe, forecasts further weakness in the economy in 1993, with recovery becoming meaningful only in 1995. By that time, oil consumption would be down to 5.4 mmb/d, about 30 percent below 1990. After that the forecast for oil consumption and economic growth are pretty much parallel; by 2010 both GDP and oil consumption would be about 10 percent higher than in 1990.[16] Even this forecast, which is more pessimistic than most, reflects surprisingly little change in oil intensity, perhaps because it is doubtful about prospects for economic reform.

Key Oil Supply Variables

Journalistic fears to the contrary, the world shows no signs of running out of oil in the foreseeable future. Nor is supply likely to be tight any time soon. In 1960 proven or identified reserves of conventional oil amounted to 256 billion barrels, enough to last the world for twelve years at the current annual level of production (equal for practical purposes to consumption). Thirty years later, after additional production of 533 billion barrels, the world's proven oil reserves had increased almost four times to 1 trillion barrels, enough to support forty-five years of current production. Between 1960 and 1990, oil drilling and recovery technology steadily improved, finding costs came down, and crude oil prices rose, all of which increased exploration, resulted in new discoveries, caused existing reservoirs to be evaluated upward, and made areas that

16. Charles Movit, "Developments in the Energy Balances of the Former Soviet Republics to 2015," in *Soviet Energy Outlook* (Washington: PlanEcon, March 1992), pp. 159–80.

were once considered submarginal newly attractive and identifiable as economic reserves.[17]

Analysts from the U.S. Geological Survey, which systematically examines data on a worldwide basis under its World Energy Resources Program, estimate that 304 to 1,047 billion barrels of conventional oil are still to be discovered, with a mean probability calculated at 605 billion barrels. Natural gas liquids might add 66 billion barrels to identified oil and 84 billion barrels to undiscovered resources, assuming that prices would not break out of their historic high of about $50 a barrel in 1989 dollars and that no revolutionary improvements in oil technology would occur.[18] Changes in either assumption could cause significant upward revisions in these estimates as a result of such conditions as new drilling and finding methods and enhanced oil recovery from existing reservoirs.

Remember that production would not go on indefinitely at a given level and then stop, as some scenarios featuring fears of rapidly diminishing resources imply. Should supply costs and prices rise, even gradually, demand would fall off and thereby extend the life of existing reserves.[19]

---

17. BP (British Petroleum) defines proven reserves as "those quantities which geological and engineering information indicate with reasonable certainty can be recovered in the future from known reservoirs under existing economic and operating conditions." See *BP Statistical Review of World Energy, June 1992* (London: British Petroleum Company, 1992), p. 2. "Identified reserves" as used by the U.S. Geological Survey is a somewhat different concept, with stress on reserves that could support production now. Nevertheless, the totals reported under the two definitions are not far apart; for 1991 BP reports proven reserves to be 1,001 billion barrels, while Masters, Root, and Attanasi (of the U.S. Geological Survey) estimate identified reserves to be 1,053 billion barrels for 1990. Allocations among countries under the two definitions show a wider difference. For this reason, their survey data are shown in table 2-4. See Charles D. Masters, David H. Root, and Emil D. Attanasi, "World Resources of Crude Oil and Natural Gas," in *Proceedings of the Thirteenth World Petroleum Congress* (Chichester: John Wiley & Sons, 1991), p. 53, table 1, p. 55.

18. Masters, Root, and Attanasi, "World Resources;" and C. D. Masters, D. H. Root, and E. D. Attanasi, "Resource Constraints in Petroleum Production Potential," *Science*, July 12, 1991, p. 148.

19. Experience after the oil shocks demonstrated that the long-term price elasticity of oil, even when used in transportation, is on the order of -0.5. *OECD Economic Outlook*, no. 48, p. 45.

Finally, a vast quantity of nonconventional recoverable oil—conservatively, on the order of 1 trillion barrels—is locked in oil shales in the United States, tar sands in Canada, and the Orinoco extra-heavy oil belt in Venezuela. Canada and Venezuela produce small quantities now, which are expected to grow moderately in the future. Should prices look sustainable, say at $30 a barrel, half again higher than at present, these resources, notably, the Venezuelan extra-heavy oil could be exploited on a much larger scale, even after allowing for higher capital costs and the investments needed to contain environmental damage.

To be sure, some signs are disquieting. Contrary to expectations, the drilling boom after the oil shocks of the 1970s resulted in numerous discoveries but did not unearth major new basins, defined as containing 20 billion barrels or more of crude oil, and analysts at the U.S. Geological Survey hazard the view they are not to be found.[20] Some experts also expect that at some point in the next two decades, additions to reserves will fall behind production, leading to the beginning of a long decline in the total identified reserves in the world. Nonetheless, current estimates show that enough oil exists in the world to sustain substantial production through the twenty-first century and probably beyond, certainly enough for the orderly energy transition that technological advances should make possible.

The problem for the issue of oil security is not the size of these resources but their distribution. Three-fourths of identified reserves are in OPEC countries and two-thirds are in the Middle East (table 2-4). The distribution of estimated undiscovered oil does not change matters much. Maldistribution appears in other commodities, without oligopolistic or political consequences and, therefore, is not viewed as a problem. Oil, as already discussed, is different, and later in this volume, the political outlook in the Middle East as it may affect the supply

20. Masters, Root, and Attanasi, "Resource Constraints," pp. 146–52.

Table 2-4. *World Oil Resources*
Billions of barrels

Region	Production 1991	Identified reserves 1990	Ratio, reserves to production	Mean probability for undiscovered oil[a]
OECD				
United States	3.29	45	14:1	49
Canada	0.72	13	18:1	28
United Kingdom	0.69	18	26:1	11
Norway	0.70	16	23:1	9
Other	0.44	5	11:1	7
Total	5.84	97	17:1	104
Former Soviet Union	3.81	83	22:1	122
Other non-OPEC countries				
Mexico	1.08	45	42:1	37
China	1.03	31	30:1	45
Other	2.62	63	24:1	87
Total	4.73	139	29:1	169
OPEC				
Saudi Arabia	3.16	267	84:1	62
Iraq	0.08	106	103:1[b]	45
Kuwait	0.06	93	140:1[b]	4
Iran	1.19	74	62:1	22
United Arab Emirates	0.96	66	69:1	7
Venezuela	0.97	49	51:1	35
Other	2.64	80	30:1	35
Total	9.06	735	81:1	210
Total world	23.44	1,053	45:1	605

Sources: *BP Statistical Review of World Energy, June 1992*, p. 5; and Charles D. Masters, David H. Root, and Emil D. Attanasi, "World Resources of Crude Oil and Natural Gas," in *Proceedings of the Thirteenth World Petroleum Congress* (Chichester: John Wiley & Sons, 1991), pp. 52–53.

a. There is a 95 percent probability for total world oil resources of 304 billion barrels, and a 5 percent probability of 1,047 billion barrels.

b. Calculated on basis of production in 1989.

of oil from that area and its potential for bringing disorder to world markets is assessed.

For present purposes, assume that during the next twenty years OPEC will continue to act as residual supplier to the world oil market. It will seek as best it can to extract economic rent from its oligopolistic position but will always confront two limitations: the need to reconcile decisions to increase prices with the fear of losing markets as a result of improvements in efficiency and substitution effects spurred by higher prices; and the need to reconcile differences in the pricing strategies among its members arising from the great variations in their oil resources, finances, and political interests. We start, therefore, by assessing the supply prospects in non-OPEC countries, which at a given level of world consumption will determine the call on, or demand for, OPEC oil and then ask what might be said about investment in the production capacity of OPEC countries to meet those requirements.

Production in the OECD

OECD countries, whose production by definition is the most secure, now account for one-fourth of the world's supply of oil. That proportion is certain to decline. The three major sources—the United States, the United Kingdom, and Norwegian sectors of the North Sea and Canada—constitute a resource base close to, or already, mature.[21]

In the United States, still the second largest producing country in the world, crude oil production peaked in 1970. Even with much higher prices and the addition of Alaska, it was unable to regain that level in the first half of the 1980s. The collapse of prices in 1986 caused the loss of 1 mmb/d from marginal wells, and a slow-paced decline is likely to continue. During the past six years returns from exploration and production have been low, and cash flows have only partially been reinvested. A sample of six major companies shows that only 31 percent of their capital spending on exploration in 1991 occurred in the United States compared with double that proportion in 1985.[22] Environmental restrictions on exploration in Alaska, the eastern Gulf of Mexico, offshore the East Coast, and on production in offshore California are also a constraint.

---

21. Ibid., p. 149. The authors state, "When prices, over the range of historical experience, limit entry of the industry into frontier or high-cost areas or prevent sufficient drilling to maintain production, the industry is mature."

22. William L. Randol and Victoria L. Hallstrom, "Chronicling the Exodus of Capital from the Domestic Oil Industry," United States Equity Research: International Oil (New York: Salomon Brothers, July 1992), p. 5, fig. 3.

The Energy Information Administration forecasts that crude oil production will decline 1.7 mmb/d between 1990 and 2010, even with gradually rising prices, mitigated in part by a moderate increase in liquid fuels from natural gas (NGL).[23]

An increase in North Sea production of about 2 mmb/d is expected over the next five years or so. The Norwegian government has scrapped its capacity restrictions, and numerous recent discoveries of small fields will be coming into production in the U.K. sector. Total North Sea production is likely to peak at 6 mmb/d at the turn of the century and then decline, with most expectations being that by 2010 production will be at roughly 4 mmb/d, about the same as in 1990. That course of development is consistent with the comparatively meager resources believed to remain.

Canadian production has remained fairly stable at about 2 mmb/d for some time and could well be at that level or somewhat higher in two decades. Production of conventional oil may slip even though some promising frontiers are awaiting development, but nonconventional oil makes the difference: by 2010 oil production from tar sands could be 500,000 barrels a day.

In sum, by the beginning of the next century, the OECD oil basin, as a whole, could have reached the point at which the current level of production would no longer be sustainable.[24] A slow decline would then begin, with production by 2010 falling somewhat below the present level, with the prospect of further declines.

Production in the Former Soviet Union

Eventually production in the former Soviet Union will recover from its present free fall but stabilize well below its

23. Energy Information Administration, *Annual Energy Outlook 1993 with Projections to 2010*, DOE/EIA-0383 (93) (Department of Energy, January 1993), p. 38, table 8.

24. Joseph P. Riva, Jr., "World Oil Distribution," *CRS Review*, vol. 12 (March-April 1991) (Washington: Congressional Research Service), pp. 5–8.

historic peaks. Production first exceeded 12 mmb/d in the early 1980s with the accelerated exploitation of the Siberian oil resources. It reached a peak of 12.6 mmb/d in 1988 and then began to decline sharply because of investment cutbacks and equipment shortages. In 1990 civil strife in Azerbaijan, the center of the oil service industry, shut down for several months plants that produce a major share of the pipes, drill bits, and other service goods for the entire oil industry in the former Soviet Union. Imports of foreign equipment and supplies could not fill much of the gap because of the shortage of foreign exchange. As a consequence, maintenance, production, and exploration activities have been penalized, many wells are closed in, and not enough new wells are being drilled to develop new reserves. Add to this the effect of perverse price incentives and squabbles among local, regional, and national officials and the results are predictable. Production fell to 10.4 mmb/d in 1991, 9.0 mmb/d in 1992, and is expected to decline to 8.0 mmb/d in 1993.[25]

Setting aside the current disorganization of the economy and bottlenecks in the oil service sector, widespread mismanagement of the oil fields in the 1980s is also taking its toll. In the effort to meet overly ambitious production targets, which were driven by the need for foreign exchange, the authorities, typically, sacrificed ultimate recovery of resources in favor of maximizing current output. For example, reliance on imprecise drilling technology, waterflooding wells, and flaring gas extracted much less oil than more modern methods and ended up waterlogging the Siberian fields. A reporter in April 1991 described the situation this way:

> The proportion of water among the liquid that comes to the surface in western Siberia has risen from 38% in 1982 to 67% in 1989. In the Tyumen area, a third of all oil wells are not now producing oil. Many are frozen over:

25. International Energy Agency, "End-January Oil Market Report" (Paris, March 8, 1993), p. 25, table 1.

they will have to be expensively redrilled. . . . Given such dreadful technology, the Soviet Union will need to double the number of wells it drills in the next five years simply to maintain today's oil output.[26]

Recovery will depend on new discoveries, but they will not come easily. Known reserves have been depleted more rapidly than anticipated, and new discoveries are more difficult and costly to develop. Current forecasts suggest that production will bottom out by around mid-decade at 8 mmb/d and then gradually recover to 11 mmb/d by 2010. Analysts at the U.S. Geological Survey have a decidedly more bearish view, estimating that by 2010 production will reach only 7.5 mmb/d and that proven reserves will be down by one-quarter from 1991 levels.[27]

Clearly, projections for this sector of the world oil supply are subject to more than the usual uncertainties. An important consideration is the role of foreign investment and technology. Until now the evidence has been mixed. In 1992 Kazakhstan concluded four major long-term agreements that promise $40 billion in foreign investment over the next forty years, including critically needed technology for oil recovery.[28] However, in Russia, which is believed to contain 90 percent of the oil in the former USSR, negotiations with foreign companies are entangled in domestic political and ideological squabbles and a cumbersome bureaucracy.[29] At some point, the need for foreign

26. "The Soviet Energy Industry: Powerless," *Economist*, April 13, 1991, p. 68. Another useful account of the oil industry's troubles appears in Bernard J. Picchi and Ann L. Kohler, "The Soviet Petroleum Industry—A Titan Struggles" (New York: Salomon Brothers, March 1991). Also, IMF, World Bank, OECD, and EBRD, *The Economy of the USSR*, pp. 40–43.

27. Masters, Root, and Attanasi, "Resource Constraints," p. 150.

28. "Kazakhstan's Oil Industry: Tomorrow's Gusher," *Economist*, July 25, 1992, p. 72. A definitive agreement for the largest of these ventures was finally signed by the Republic of Kazakhstan and the Chevron Corporation in April 1993. Both parties described it as a $20 billion joint venture estimated to produce 700,000 barrels a day by 2010. Kathryn Jones, "Kazakhstan and Chevron Start Venture," *New York Times*, April 7, 1993, pp. D1–D2.

29. A vivid description of the obstacles currently confronting present and prospective foreign investors appears in Ann Imse, "American Know-How and Russian Oil," *New York Times Magazine*, March 7, 1993, pp. 28–31, 57, 66.

exchange and the pressure of domestic oil demand should force the pace. Moreover, substantial foreign assistance, both multilateral and bilateral, is aimed at easing domestic bottlenecks and facilitating foreign investment in the Russian oil industry. Similarly, domestic price distortions in the oil industry should lessen, perhaps slowly but inexorably, except in the unlikely event that a new coup were to succeed in restoring the Soviet past. Nonetheless, the resource base, large as it may be, can hardly support a return to the production heights of the 1980s, if fields are exploited at maximum efficient operating rates.

Production in Non-OPEC Developing Countries

Continued oil production gains are in prospect for the non-OPEC developing countries but at a much slower pace than during the past two decades. This group includes two major producers, Mexico and China (which account for almost half of the total), seventeen countries each of whose production has or is expected to reach at least 200,000 barrels a day, and a number of small producers whose combined production is likely to stay at about 2 mmb/d. Since 1973, production from this group of countries has increased an average of 5.5 percent a year; forecasters have consistently underestimated its potential.

Mexico's production rose rapidly to 3 mmb/d in 1982 and has pretty much stayed there even though the country's identified reserves are about equal to those of the United States and exploitable at lower average cost. Oil exploration and development are constitutionally limited to the state oil company (Pemex) to the exclusion of foreign and private domestic investors. Pemex has been riddled by problems in the past, has been a principal target of the Salinas government's domestic reform program, and has been forced to cut its investment program over the medium term. In the oil sector, the government's overtures to foreign investment have been limited to providers of oil services. In 1991 the U.S. Ex-Im Bank extended a $1.3 billion loan guarantee program for U.S. companies to participate in such activities. Innovative joint venture arrangements are under way that may widen the participation of foreign companies without breaching the constitutional limitation. The North American Free Trade Agreement could spur this trend. Without greater foreign company investment or the development of an independent and private domestic oil industry, Mexico's production prospects look flat; with either or both, the development of reserves could begin to increase once again and production could reach a minimum of 4 mmb/d by 2010.

China's production, in 1991 at 2.8 mmb/d, increased 7 percent a year between 1973 and 1985 and about 2 percent annually in the late 1980s. The government target now calls for production to increase by only 1 percent a year through 1995. Offshore and onshore oil fields in south China, the only areas where foreign companies were permitted to participate, have been disappointing. The main producing areas in the northeast and eastern regions have lost some steam; outside observers believe that lifting the rate of increase will require more drilling and development expenditures there than the government has committed. New oil-bearing areas in northern and western China are believed to have the greatest potential. Notably, a recent discovery in the Tarim Basin of northwestern China could turn out to be a supergiant field, although data are not yet available for a definitive assessment. The U.S. Geological Survey describes it as a low-probability, high-volume occurrence. In any event, exploitation of the field would require construction of a lengthy and costly pipeline to the nearest port. In a noteworthy change of policy, the Chinese authorities announced in February 1993 that foreign oil companies could bid for rights to explore for and develop oil fields in specified sections of the new regions.[30]

The remaining countries in the group are characterized generally by a newly strengthened interest in reaching explo-

---

30. Masters, Root, and Attanasi, "World Resources;" and James McGregor, "China to Open Oil Frontier to Foreigners," *Wall Street Journal*, February 18, 1993, p. A13.

ration and development agreements with foreign companies, stimulated in part by, and tending to offset, the depressing production effects of the collapse of oil prices in 1986. By and large the substantial flow of exploration funds moving out of the United States has been going to countries in this group. Some discoveries have occurred, most notably in Colombia, Yemen, Syria, Papua New Guinea, and Tunisia. Brazil has uncovered potentially important possibilities offshore and a high volume of exploration activity may keep production in Egypt just under 1 mmb/d. India and Malaysia's production, however, seems likely to decline over the medium term. The consensus is that production from this group of countries as a whole will rise substantially through the end of the decade as the new discoveries reach peak output and then decline slowly to around present levels by 2010.

OPEC Capacity Policies

OPEC's willingness to invest sufficiently in capacity to meet a growing world demand for oil is a question that its individual members will determine; as an institution, the cartel has never been entrusted with the responsibility for systematically addressing this issue. Further, the answer will depend mainly on the individual decisions of the five large Persian Gulf producers and Venezuela. They alone have the resources that could shape world markets over the next two decades, the more so beyond that. It is not by accident that in 1960 these countries, excluding the United Arab Emirates (UAE) which then was barely beginning to produce oil, were the founding members of OPEC.

The other seven members now account only for about one-fourth of OPEC's production. On the basis of present plans, their collective capacity can be expected to grow moderately through the turn of the century. In view of resource limitations, however, by 2010 production is not likely to be larger than it is now.

History has some useful reminders of how the major producing countries in OPEC addressed the question of capacity over the past thirty years. Between 1967 and 1973, the international oil companies through their concession agreements had at least nominal control over production decisions in the Middle East. They increased capacity steadily to meet demand that was more than doubling. In doing so, the companies had to reconcile the competing demands of the king of Saudi Arabia and the shah of Iran for production increases in their respective countries, while having to worry about the further weakening of oil prices that could result. OPEC countries supplied three-fourths of the increase in world oil consumption during the decade, of which Saudi Arabia and Iran accounted for more than half. No evidence from this period of booming oil markets suggests that OPEC countries saw any connection between decisions on capacity and price. Instead, their big concerns were to extract a larger share of the concessionaires' take, where their demands were measured in cents not dollars per barrel, and to increase revenue as well from higher production.

By the time of the first oil shock in 1973, OPEC's sustainable operating capacity had increased to 34 mmb/d as against actual production of 31 mmb/d. During the following five years, capacity remained fairly constant as demand for OPEC's oil dropped sharply in response to the price jump and then recovered to the pre-shock level by 1979. Prices fell during this interval by about 10 percent in real terms, principally because Saudi Arabia played a moderating role. The Saudis constantly threatened to use their surplus capacity to restrain OPEC pricing decisions and actually did so in 1977 when it temporarily split with the other members to bring a two-tier pricing system into being.

During the second oil shock, set off by the Iranian revolution in 1978 and then fostered by the Iran-Iraq war beginning in September 1980, Saudi Arabia again sought to restrain price increases by producing at or near capacity for almost three years. It failed to prevent price escalation because the market effects of its additional production could not offset output reductions by other OPEC members, who were determined to keep prices high.

Between 1979 and 1985, under the pressure of high oil prices on world consumption and the steady increase in non-OPEC production, the demand for OPEC oil fell by almost one-half. OPEC capacity also languished because of the damage to Iran-Iraq oil installations in their protracted war, inadequate maintenance as excess capacity grew, and a secular decline in Venezuelan output because of resource depletion, which subsequent new discoveries will reverse. Saudi Arabia, which early in the decade seemed intent on increasing its capacity by 2-3 mmb/d, abandoned its plans. By 1985 OPEC operating capacity had fallen to between 27 and 28 mmb/d compared with a production low of 15 mmb/d in the third quarter, leaving the largest residue of surplus capacity in the history of the oil industry. When Iraq's invasion of Kuwait cut off supplies from both countries and touched off a price spike, the other OPEC countries, principally Saudi Arabia, brought all available idle and mothballed capacity into production, at comparatively moderate investment cost, to restore balance in the market.

Now, for the first time since the 1960s, the OPEC countries must seriously address the issue of when and by how much to increase production capacity. That world consumption since 1985 has been increasing on average by 1 mmb/d a year, with OPEC the main beneficiary, brings the question to the fore. Available excess capacity is now small. With Iraq returning to production, a cushion of 3-4 mmb/d would emerge, not excessive by normal market standards. Furthermore, continued decline in production from the former Soviet Union is possible, the U.S. production decline could quicken, and economic recovery in the OECD would call for more OPEC oil. As noted earlier, the smaller OPEC producers are planning to increase capacity, some with the newly invited participation of foreign companies. In making their decisions, the six critical countries must grapple with economic, financial, and political concerns.

What Saudi Arabia does will be key to OPEC's course of

action, greatly influencing the evolution of the world energy system. Its linked decisions on production capacity and price rest on well-understood considerations. The country's huge oil reserves give it a strong interest in encouraging the long-term growth of demand, which it is less prone than others to sacrifice for short-term windfalls. Its financial surplus, now much reduced but bound to grow again, gives it a tangible stake in the health and financial stability of the world economy. And the government's conservatism puts it in the camp of preserving stability in the world. Saudi leaders have consistently stated that oil shocks could put all of these interests at risk.[31] On top of that is Saudi Arabia's longstanding and wide-ranging relationship with the United States, which long since came to represent something close to an implicit U.S. guarantee of Saudi security. That chit was called and honored by the United States and its allies when Iraq invaded Kuwait. The outcome of the Persian Gulf War has given Saudi Arabia greater freedom to pursue its oil interests with little to fear from retaliation by its price-hawkish neighbors in the region. To advance those interests, experience has demonstrated to the government that the existence of substantial excess Saudi capacity is essential. Without it, the Saudis cannot restrain a tight market, a situation that opens the way to panic and price shocks. At one time the traditionally cautious Saudi authorities seemed to want to avoid building spare capacity. They feared being caught between the threats of aggressive OPEC members if they used it and the importuning of the United States and other industrial countries to do just that. The damaging economic and oil mar-

31. The government's history of counseling price moderation seems ironic in view of the fact that Saudi Arabia was the critical force behind the Arab oil embargo in 1973, which set off the first oil shock. Saudi Arabia was also the first country, however, to restore full production and lift the embargo when U.S. officials made that a condition for U.S. involvement in the search for a solution to the Middle East crisis. Within OPEC, moreover, the Saudis were at the low end of the price proposals advanced during the hectic OPEC meetings of October to December 1973 and warned that the oil shock that resulted would ultimately be a disaster for OPEC and the oil industry. See Ian Skeet, *OPEC: Twenty-Five Years of Prices and Politics*, Cambridge Energy Studies (Cambridge University Press, reprinted 1989), pp. 99–123.

ket effects of the first two oil shocks and the repulse of the Iraqi invasion should have buried that concern.

Kuwait permitted its capacity to decline during the 1970s and 1980s to about 2 mmb/d as it pursued an essentially restrictionist crude oil production policy. At the same time it invested heavily in refining and marketing facilities in OECD countries and saw its huge external investments as a secure, growing source of its financing needs. All that is now changed. Oil production has recovered more rapidly than had been anticipated, having reached the preinvasion level of 1.6 mmb/d in January 1993. The investment hoard, however, is severely depleted just as reconstruction and oil development costs are running very high.[32] In the circumstances, Kuwait can be expected to increase oil production capacity substantially in the future and support Saudi Arabia's oil price strategy even more closely than in the past. The UAE, another Saudi Arabian oil ally, should follow much the same policies as Kuwait, even though it did not suffer damage from the war.

Iraq had been intent on expanding capacity before the invasion, reportedly planning an early increase in exports to 5 mmb/d from under 3 mmb/d in 1989. Some of this claim was evidently militant swaggering, but some large recent discoveries were near the development stage when Iraq made its move. Disuse has probably brought neglect to the oil fields, which means deferral of expansion plans when Iraqi oil exports resume, with or without Saddam Hussein. No matter who is in charge when Iraq rejoins the community of nations, the country can be counted on to push for expansion of its capacity in support of its chronic battle for market share in OPEC, to finance its huge reconstruction requirements, and to provide the basis for reviving its prestige and influence in the region.

Iran's capacity, now about 4 mmb/d, has been cut by one-third because of neglect of oil reservoirs after the revolution,

---

32. "Into a Black Hole: Kuwait," *Economist*, January 30, 1993, p. 40, places the current Kuwaiti investment holdings at $30 billion, down from $100 billion before the invasion.

damage from the war with Iraq, and low investment in exploration and development. Capacity can be expanded considerably through the repair of damaged fields, enhanced recovery techniques, and significant recent discoveries. Additional oil production is essential to the Rafsanjani government's plans to resume economic development on a large scale. As part of its opening to the West, Iran has invited French companies to assist in its oil expansion program. The government also seeks to recover some influence in OPEC, going back to the early 1970s when the shah and Saudi Arabia vied for leadership of the organization. To this end, greater production capacity is important.

Expansion of Venezuela's capacity follows from its new discoveries of light oil and continuing investments in the exploitation of extra heavy crude. The state oil company has ambitious investment plans. In a reversal of policy, moreover, foreign companies, for the first time since Venezuela nationalized its oil fields in 1976, have been invited to explore for crude oil and develop its production. Venezuela has been an organization stalwart ever since it initiated the meetings that led to the founding of OPEC. Over the years, it consistently favored high oil prices and avoided, more or less, cheating on its production quotas. These views were never a barrier to expanding its capacity within the limits of funds and oil resources. Moreover, it has invested heavily in marketing and refining facilities in the United States and Europe, which are ready markets for its crude oil.

In sum, each of the major OPEC producers intends to increase production capacity in the light of what looks like an expanding market for OPEC oil over the indefinite future. These intentions must have been reinforced by Saudi Arabia's position in OPEC meetings suggesting that production quotas should be based mainly on capacity, divorced once and for all from arguments about comparative need. In any event, expansion is clearly under way. One analyst, reviewing official statements over the past few years and tempering them by judgment, estimates that the Persian Gulf OPEC members could have 8–9 mmb/d of additional capacity in place by 1997, say, more conservatively, by the end of the decade, with another 1 mmb/d coming from other OPEC members.[33] That would be well within their resources and consistent with their motivations.

Outlook for the World Oil Market

Medium- and long-term forecasts of the oil market are based on economic models incorporating the relationships underlying demand, supply, and price, complicated by interpretations of OPEC's effectiveness as a cartel. Some are fairly simple; others elaborate and econometric. A GDP growth rate is assumed, usually taken from independent forecasts. Generally an oil price path is projected from the past behavior of price in relation to OPEC excess capacity, which in turn requires judgments about the growth of that capacity. Oil demand rests on assumptions about oil price, economic growth, and the pace of technological advance. Estimated non-OPEC production depends on reliable information on finding costs, which is only selectively available, and on attitudes of the host country toward foreign investment.

In the early days of the OPEC oil shock decade, these forecasts proved to be far off the mark, consistently predicting much higher prices than actually occurred. After the collapse of oil prices in 1986 and as more data accumulated, predictions have been coming closer to reality. Even now, however, wide variations appear, depending on the modeler's interpretation of one or another of the underlying oil relationships. Demand projections are the most striking case in point. The Energy Modeling Forum covers the work of eleven comprehensive oil models. Starting from common assumptions about economic growth and oil prices, projections for world demand for oil in

---

33. Adam E. Sieminski, "Oil Production Outside the Gulf, 1992-1997" (Washington: NatWest Investment Banking Group, April 1992), pp. 1–2.

the year 2000 show a difference of 30 mmb/d between the low and the high estimates![34]

With these caveats in mind, we propose using the international oil market outlook developed in 1992 by the U.S. Energy Information Administration as a starting point for examining oil security. The EIA projection is comprehensive and detailed, covers the period addressed in this book, contains an uncertainty range for its projections, and is reviewed each year with the benefit of new data, new announcements of official plans, and a lengthy learning experience. For each year's projection, existing country policies on energy, oil, and the environment are assumed to remain unchanged. The 1991 results are summarized in table 2-5. The base case shown in table 2-5 and used for our analysis "is not presented as a most likely case, but simply as a representative case."[35]

The key findings of this forecast follow. Spurred by world economic growth assumed to average 3 percent a year through 2010, energy consumption grows by one-half, and oil consumption by one-third the growth rate. For the remainder of the 1990s, non-OPEC production meets about one-half the growth in oil demand and OPEC supplies the rest. In the first decade of the twenty-first century, non-OPEC production declines very slowly while OPEC is projected to supply the entire increase in oil demand—annually about 1 mmb/d — plus a bit more. OPEC capacity is shown to grow by enough to meet this increase in world demand and to maintain a cushion that would more than cover cyclical and seasonal variations. Oil prices in 1990 dollars are projected to increase from $18.40 in 1991 to $33.40 in 2010, on average by 3.2 percent a year. The price path projection seems inconsistent with the forecast OPEC share of the market: oil prices are projected to rise 3.2 percent

34. Energy Modeling Forum, *International Oil Supplies*, pp. 6–9, explains how this and other differences can come about.

35. Energy Information Administration, *International Energy Outlook 1992*, DOE/EIA-0484 (92) (Department of Energy, April 1992). The Energy Information Administration (EIA) was established as an independent statistical and analytical agency in the Department of Energy. See pp. viii–ix.

Table 2-5. *World Oil Outlook, 1991–2010*[a]

Million barrels per day unless otherwise noted

Item	1991 (actual)[b]	Base case		
		1995	2000	2010
Consumption				
OECD	37.2	40.3	41.6	43.6
Former Soviet Union	8.4	6.5	7.5	8.9
Developing countries[c]	19.9	22.8	25.9	31.8
Total world	65.5	69.6	75.0	84.3
Uncertainty range	...	67–72	71–78	79–90
Production				
OECD	16.0	18.0	17.2	15.0
Former Soviet Union	10.4	8.5	9.5	11.0
Non-OPEC developing countries	13.0	15.7	17.4	16.1
OPEC	24.8	27.1	30.9	41.9
Total world	64.2	69.3	75.0	84.0
Estimated OPEC excess capacity	4–5[d]	3–4	3–5	3–5
Oil price in 1990 dollars per barrel[e] (base case)	18.40	20.80	26.40	33.40
Uncertainty range ($)	...	16.00–25.30	17.90–31.80	22.60–40.20
Annual growth of world GDP (%)	...	2.8[f]		3.0[f]

Sources: Energy Information Administration, *International Energy Outlook 1992*, DOE/EIA-0484 (92) (Department of Energy, April 1992), tables 1–5, pp. 9–13; tables 2-1, 2-2; and authors' estimates.

a. Includes crude oil, natural gas liquids, condensates, and refinery gains.

b. From tables 2-1 and 2-2.

c. Developing countries include China, other centrally planned economies, and Eastern Europe.

d. Assumes Iraq and Kuwait producing at 1989 levels.

e. U.S. refiner acquisition cost of imported crude oil, which is comparable to the average OECD import price.

f. Growth of 2.8 percent applies to the period between 1990 and 2000; 3.0 percent to the period between 2000 and 2010.

a year through 1995, 5 percent a year in the following five years, and only 2.4 percent a year between 2000 and 2010 when OPEC is expected to capture all the expected increase in demand. The uncertainty range, however, is wide enough to cover situations at the low end in which real prices barely rise through 2010 and at the high end in which they increase by more than 4 percent a year.[36]

36. Energy Information Administration, *International Energy Outlook 1992*, pp. vii–16. In the 1993 edition of its annual energy forecast for the United States, EIA has greatly reduced its world oil price assumptions. They now range in 1991

We would modify these findings as follows:

—The base projection of oil demand probably understates prospective improvements in the efficiency of oil use in the developing countries and in the former Soviet Union and therefore overstates the growth of world oil consumption.

—Oil prices over the next few years should not increase at all, in real terms, and could temporarily go down, because of the likely need at some point to accommodate the resumption of Iraqi oil exports.

—Until the turn of the century, expansion of OPEC capacity will probably outpace the increase in the call on OPEC oil, again indicating price restraint, not large price increases.

—More generally, from now to 2010 the price of oil will depend largely on Saudi Arabia's position, supported by its Gulf Cooperation Council allies. The Saudis can confidently prevent a price collapse or even a significant erosion of prices because, in contrast to the years preceding the 1986 price collapse, OPEC faces an increasing, not a sharply declining, demand for its oil and because the other cartel members now know from experience that the Saudis will abandon OPEC production quotas if the others flout them excessively. At the same time, Saudi interests in controlling OPEC pricing policies and in insuring a steady increase in its own oil revenues mean encouraging demand, which requires comparatively stable or only moderately higher real prices.

In any event, even with oil prices nearer the lower end of the uncertainty range shown in table 2-5, that is, with relatively constant real oil prices, which is a more plausible forecast, the oil security implications can be summarized as follows:

—Oil would account for a moderately smaller share of

---

dollars from $18.70 a barrel in 1991 to $22.90 in 2000 and $29.30 in 2010. For 2010, the assumed price is $5.46 a barrel below that shown in table 2-5. The average price increase through 2010 is 2.4 percent a year. Energy Information Administration, *Assumptions for the Annual Energy Outlook 1993*, DOE/EIA-0527 (93) (Department of Energy, January 1993), p. 4. These new world oil price assumptions presumably will be used for the Energy Information Administration, *International Energy Outlook 1993* (Department of Energy, forthcoming).

world primary energy consumption, declining from 39 percent in 1991 to 36 percent in 2010.

—The OECD share of world oil consumption would continue to fall, from 57 percent today to around one-half in 2010, while the share of developing countries will rise to 38 percent, up by one-third since 1991.

—Oil would be of declining importance in the OECD economy, with oil costs falling from 1.9 percent of total output in 1990 to 1.4 percent in 2010.

—Oil developments in the former Soviet Union should not be a source of serious new disturbances to the world oil market. Net exports, which peaked at close to 4 mmb/d in 1988 declined steadily since then as domestic consumption fell less rapidly than production. They were down to 2 mmb/d in 1992, with a further modest decline expected in 1993. The fall in net exports thus far has been offset by OPEC's use of its excess capacity and has already been absorbed by the market. When recovery of the oil industry and the economy are in place, increases in oil consumption are likely to lag behind increases in oil production because of expected improvements in the efficiency of oil use. Net exports could stabilize in the neighborhood of 2-3 mmb/d.

—By 2010 OPEC would be supplying half of world oil production compared with 39 percent in 1991. This amount is about the same proportion that OPEC supplied in 1973, at the time of the first oil shock. The Persian Gulf countries alone would account for 28–30 mmb/d, half again as much as they supplied in 1989, and 38 percent of the world total (figure 2-2).

—OECD oil import dependence would rise from 57 percent in 1991 to 66 percent in 2010, with further increases in store after that time. For the United States, typically more preoccupied with that concern than other OECD countries, oil import dependence would grow to 59 percent from 46 percent in 1991, with imported oil for the first time exceeding domestic production as early as the mid-1990s.

In sum, these findings suggest that the OECD economies

Figure 2-2. *OPEC Oil as a Percentage of World Oil Production, 1973–91, Selected Years, and Projections for 2000 and 2010*

Percentage

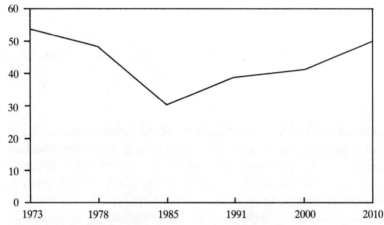

Sources: *BP Statistical Review of World Energy, June 1992*, p. 5; *BP Statistical Review of World Energy, June 1985*, p. 5; *BP Statistical Review of World Energy, June 1984*, p. 5; and table 2-5.

specifically and the world economy in general will be moving toward a position over the next two decades where they will be increasingly vulnerable, statistically, to oil supply interruptions from the Middle East. Later in this book we examine the likelihood that political developments in the region could again cause interruptions in the oil supply.

CHAPTER THREE
Environmental Uncertainties

AMONG THE MANY uncertainties about future demand for oil is the prospective international response to environmental concerns. These concerns, and the political activism they have inspired, have helped, so far only modestly, to diminish oil intensity in the OECD countries since 1973. As the 1992 United Nations Conference on the Environment and Development at Rio de Janeiro showed, the environmental movement has by no means lost momentum. Rather it seems to have gained ground where it had had the least impact, that is, in the developing world. The influence of environmentalism on the oil market over the next decades is hardly in question. Still, whether it will be a principal factor in itself for reducing oil demand is doubtful.

Environmental concerns stemming from the use of oil have two main strands. One has to do with emissions of sulphur dioxide (SO_2), carbon monoxide (CO), nitrogen oxides (NO_x), and volatile organic compounds (VOC) associated with urban smog and acid rain. These pollutants are considered to be currently or potentially controllable at less than prohibitive costs. The other category of concern is the emission of man-made, heat-trapping gases, which are putative contributions to the greenhouse or global warming effect. The largest share of these pollutants is represented by carbon dioxide (CO_2) from the combustion of coal, oil, and natural gas. Commercially viable technology to control CO_2 emissions does not now exist.

Pollutants Other Than Carbon Dioxide

Various measures have been introduced to reduce flows of this group of pollutants. Volume limits on emissions have been the favored approach, however, in the United States and Western Europe. Protocols to the 1979 Convention on Long-Range Transboundary Air Pollution have set target ceilings on SO_2, NO_x, and VOC emissions; in a few cases—for example, SO_2 in western Germany—emissions have been reduced well beyond the targets.[1] The United States has had emission standards for motor vehicles since 1967, which were tightened in 1990. Japan and the European Community states have similar standards. The state of California has set the most stringent limits anywhere. Nine other American states and the District of Columbia, together with California constituting about one-half the U.S. new car market, have expressed an intention of following California's lead. A national fuel economy target (27.5 miles per gallon of gasoline) for new cars and light trucks has been in place in the United States since 1985.[2]

Regulatory measures of this sort normally will add to consumer costs for gasoline and thus tend to reduce oil consumption. At the same time, controls on vehicle emission will tend to reduce fuel economy, if only by reason of the added weight of the control devices.[3] According to the Department of Energy, the net result of the 1990 Clean Air Act amendments will be to displace, by 2010, 600,000 to 700,000 barrels a day of oil consumption in the transportation sector (baseline 15.4

1. World Resources Institute in collaboration with the United Nations Environment Programme and United Nations Development Programme, *World Resources 1992–93* (Oxford University Press, 1992), pp. 199–201, table 24.5, p. 351.

2. National Research Council, *Automotive Fuel Economy: How Far Should We Go?* (Washington: National Academy Press, 1992), pp. 12, 71–75. The legislation instituting this requirement passed in 1975.

3. National emission standards in the United States are expected to reduce fuel economy by "roughly 0.5 mpg" (1.8 percent of the 27.5 mpg set by the corporate average fuel economy (CAFE) legislation). California's more stringent standards are estimated to reduce fuel economy by 3 percent to 4 percent. Ibid., pp. 76, 83–84.

mmb/d), mainly by way of substitution of the oxygenated additives required for reformulated gasoline.[4]

The U.S. energy strategy has much larger gains projected for the transportation sector, however. Alternative fuels—ethanol, methanol, compressed natural gas—would be promoted by requiring their use in certain vehicle fleets, by incentives to manufacturers to produce fuel-flexible cars, and by subsidies, primarily for research and development (R&D). These fuels are projected to displace another 1.5 mmb/d of oil in 2010. Hoped-for improvements in propulsion technologies, including advances in the electric car, could save a further 1.2 to 1.7 mmb/d (totaling 3.3–3.9 mmb/d).[5]

These hypothesized savings and their enabling policies are to be sought after and justified primarily on grounds of energy security. Environmental considerations can be a further justification but need not be. Ethanol from corn, for example, is not only an expensive if politically popular way to provide a larger market for the grain, but its production and distribution is both energy intensive and polluting.[6] The authors of the energy strategy volumes are careful also to warn that the actual "level and timing of the penetration of both flexible-fueled and dedicated-alternative vehicles into transportation markets is uncertain."[7]

Stationary sources of polluting emissions—electric power

4. Department of Energy, *National Energy Strategy, Technical Annex 2: Integrated Analysis Supporting the National Energy Strategy: Methodology, Assumptions and Results*, DOE/S-0086P (Washington, 1991), pp. 29, 47.

5. Ibid., pp. 47, 49–50, table B-9, p. 104, table C-9, p. 121.

6. M. A. DeLuchi, *Emissions of Greenhouse Gases from the Use of Transportation Fuels and Electricity*, vol. 1, *Main Text*, ANL/ESD/TM-22 (Argonne, Illinois: Argonne National Laboratory, November 1991), pp. 21–25, tables 3, 4.

7. Department of Energy, *National Energy Strategy, Technical Annex 2*, p. 168. And p. A–8 of Department of Energy, *National Energy Strategy: Powerful Ideas for America*, First Edition 1991/1992 (Washington, February 1991), says, "Even optimistic estimates project significant market penetration of alternative-fuel vehicles only after extended periods of high oil prices have stimulated increased development efforts, by 2010 and beyond." The energy strategy projections assume that the world price of oil, in 1990 dollars, will rise about 77 percent, over two decades, from $18.81 a barrel in 1989 to $33.33 in 2010. See p. 115, table C-3, in *National Energy Strategy, Technical Annex 2*.

Table 3-1. *U.S. Oil Inputs by User Category, 1990, 2010*
Million barrels per day

Category	1990	2010 Current policy base case	2010 Energy strategy scenario
Electric power generation	0.6	0.9	0.5
Residential and commercial	1.3	0.9	0.9
Industrial	4.1	5.3	5.6
Transportation	10.8	15.5	12.0
Total	16.8	22.6	19.0

Source: Adapted from Department of Energy, *National Energy Strategy, Technical Annex 2: Integrated Analysis Supporting the National Energy Strategy: Methodology, Assumptions and Results*, DOE/S-0086P (Washington, 1991), tables B6, B7, B8, B9, pp. 101–04, and C6, C7, C8, C9, pp. 118–21.

plants, industry, residences, and commercial establishments—currently account for about one-third of the oil used in the United States and about one-half in the rest of the OECD. These emissions can be abated by technical means such as scrubbers, by switching to cleaner fuels, primarily natural gas, or by combinations of improved technologies and cleaner fuels. Rising demand for electric power and for heating fuels can be kept in check by more efficient appliances, more intensive use of insulation, and so on. And fossil fuel emissions can be eliminated at the electricity-generating stage by resorting to nuclear, geothermal, hydro, wind, and solar power. Nuclear (and in some instances hydro) power aside, these possibilities are in the environmentalist catalogue of things to be promoted by exhortation and official actions.

All are found as elements of the U.S. national energy strategy, to be pursued by regulatory changes, tax incentives, R & D subsidies, and other inducements. The impact on fuel use, however, is expected to be registered entirely on coal and natural gas. As projected, oil use in the nontransportation sectors in 2010 would be the same under the energy strategy as in the unchanged policy base case (table 3-1). (The decline in oil use in the transportation sector, as described earlier, would come mainly from encouraging the displacement of oil by other fuels.)

With a somewhat different pattern of oil use, the savings

in oil consumption that may be achieved by the other OECD members no doubt will vary from those projected for the United States. Still, the proposed U.S. program is as comprehensive and ambitious as any likely to be considered elsewhere, and as sanguine in its stated expectations, including the hope that all of its pieces will survive legislative and fiscal scrutiny. If the gains assumed by the national energy strategy are speculative and dependent on concern for nonenvironmental factors, much the same will probably be true elsewhere in the OECD.

The rest of the world has placed lower priority on fear of environmental damage for reasons that are mostly understandable. Trend growth in oil use has been highest outside the OECD's twenty-four countries and, Eastern Europe apart, this situation promises to continue through 2010 and beyond. Reducing pollution is one of the arguments for policies designed to curb this growth, and it may be among the more powerful ones in cities—Bangkok, Sao Paulo, Mexico City—that are already heavily polluted by auto emissions. With transport fuels accounting for more than 55 percent of developing country oil use, possibilities for saving oil exist, particularly through tax and pricing policies.[8] How fast genuinely effective policies might be put in place remains to be seen. In the developing world "green" politics are still at an embryonic stage and generally no match for the push to expand economic growth.

Global Warming

At the Rio conference, the environmental threat posed by climate change—the greenhouse effect—emerged as the dominant issue on the agenda. The treaty that was agreed on

8. The World Bank finds that gasoline prices in developing countries average about $1.25 a gallon, varying from $0.40 in Venezuela to $2.60 in India. World Bank, *World Development Report 1992: Development and the Environment* (Oxford University Press for the World Bank, 1992), pp. 124–25.

at Rio did not establish the hoped-for targets for reducing emissions of the gases that supposedly cause this effect. The treaty did oblige its signatories to draft and report on their plans for emission stabilization, which guarantees that in countries with green constituencies the issue will not be allowed to go away.

Fears of global warming are based on measurements of a steady buildup of atmospheric greenhouse gases—carbon dioxide, chlorofluorocarbons (CFCs), methane, nitrous oxide, and ozone—which trap outbound radiation from the earth's surface. The increase in the atmospheric concentration of these gases, it is believed, will lead to rising temperatures all over the planet. Forecasts of consequent ecological and economic damage range from moderate to catastrophic.[9]

As noted earlier, carbon dioxide from fossil fuel combustion is considered the most important of the manmade greenhouse gases, contributing about 60 percent of global CO_2-equivalent emissions.[10] If global warming and its expected environmental damage are to be warded off, the growth rate of fossil fuel use will have to be slowed and then lowered, possibly a great deal.[11] Coal has the largest carbon content of the fossil fuels, followed by oil and natural gas, the relative proportions being roughly 100, 80, and 60.[12] Worldwide, coal and oil are

9. See, for instance, Intergovernmental Panel on Climate Change (IPCC), "First Assessment Report: Overview" (New York: World Meteorological Organization (WMO) and United Nations Environment Programme, August 1990). The panel's report represented a majority view of an international group comprising more than a hundred scientists and reviewers.

10. Committee on Science, Engineering, and Public Policy, *Policy Implications of Greenhouse Warming* (Washington: National Academy Press, 1991), p. 6, table 2.1.

11. That is necessary because the concentration of certain greenhouse gases will remain in the atmosphere for decades or centuries. Just to hold emissions at, say, 1990 levels (when greenhouse gases were being released faster than they were being removed from the atmosphere) would mean a continuing intensification of the greenhouse effect. Daniel A. Lashof and Dennis A. Tirpak, eds., U.S. Environmental Protection Agency, *Policy Options for Stabilizing Global Climate* (Hemisphere Publishing Corporation, 1990), pp. 8–13.

12. These measurements are applied to tons of carbon per million British

estimated to be about equal sources of emissions of carbon dioxide, together accounting for over 17 billion tons of a global total of 22 billion a year.[13] A strategy against the global warming phenomenon, undertaken in earnest, must thus have as a chief component, policies and actions to reduce the consumption, worldwide, of the fossil fuels.

That such a strategy will be mounted, in this or even the next decade, is open to question. Several reasons argue for skepticism.

For one, uncertainty still surrounds the greenhouse phenomenon. During the past century of increasing industrial activity, the average global temperature is said to have risen between 0.3° and 0.6° centigrade. This change could be partly or entirely attributable to greenhouse gases. Or natural climate variability could account for part or all of it.[14] According to the National Academy of Sciences, a doubling of the preindustrial atmospheric carbon dioxide level of greenhouse gases would cause a temperature rise within a range as wide as 1.5° to 4.5° centigrade. Then too, the atmospheric concentration of certain gases apparently has a significant but uncertain effect on the global temperature.[15] Initially the chlorofluorocarbons were considered to be the third most important contributor to the greenhouse effect; more recent studies suggest that CFC emissions may rather have a cooling effect that offsets their warming potential.[16] The speed and regional impact of possible

thermal units of crude oil equivalent. Peter Hoeller, Andrew Dean, and Jon Nicolaisen, *A Survey of Studies of the Costs of Reducing Greenhouse Gas Emissions*, Working Papers 89 OCDE/GD(90)8 (Paris: Organization for Economic Cooperation and Development, December 1990), p. 37, table 5.

13. World Resources Institute in collaboration with United Nations Environment Programme and United Nations Development Programme, *World Resources 1992–93*, table 24.1, p. 346.

14. Committee on Science, Engineering, and Public Policy, *Policy Implications of Greenhouse Warming*, p. 2.

15. Lashof and Tirpak, *Policy Options for Stabilizing Global Climate*, pp. 13, 102.

16. Committee on Science, Engineering, and Public Policy, *Policy Implications of Greenhouse Warming*, p. 20.

climate change are also uncertain, as are the probable effects on human activity.[17]

Again, adaptation to climate change is at least a partial alternative to preventive measures. The global warming effect will in any case be felt gradually, over decades. Planning and research to better cope with its expected adverse consequences can be offered as prudent and inexpensive policies at this early point.[18] And the National Academy of Sciences cites several "relatively inexpensive" geoengineering options to halt or mitigate the greenhouse effect—for example, the creation of stratospheric dust to reflect sunlight—that may be feasible in the future.[19]

If nevertheless, the decision were made to undertake a specifically contra-greenhouse program, ways and means would have to be addressed. Virtually all of the economists who have looked at the problem have embraced a tax on carbon emissions, or a system of internationally tradeable emission permits, as cost-effective ways to achieve reductions in carbon dioxide emission. A carbon tax, besides being a convenient device for economic models, has all the virtues commonly assigned to market-oriented policies. Its prospects, however, are open to doubt.

The size, incidence, and macroeconomic implications of a tax based on carbon emissions raise problems that politicians will not be able to dodge.

Modeling results thus far are not likely to inspire great confidence. Perhaps inevitably in view of the assumptions specified or the issues emphasized, they show a wide array of tax rates needed to stabilize or reduce emissions in the decades ahead. Thus an OECD review in 1992 of six global models that

17. World Bank, *World Development Report 1992*, pp. 61–62.

18. The report to the United Nations on the U.S. climate change approach sensibly emphasizes adaptation programs. Department of State, "U.S. Views on Global Climate Change," (Washington, April 1992).

19. Committee on Science, Engineering, and Public Policy, *Policy Implications of Greenhouse Warming*, pp. 58–60.

used common assumptions about population and output growth found that a target of 2 percent carbon dioxide reduction from 1990 levels by the year 2020 for the OECD countries (other than the United States) would require eventual tax rates of $233 to $342 a ton of carbon (in 1990 dollars $28.40 to $41.69 a barrel of oil). For the United States, the range would be $324 to $354. For China the carbon tax spread was $26 to $320, for the former Soviet Union $69 to $322.[20] An International Energy Agency (IEA) model, taking the same carbon reduction target for the period 1990 to 2005, would call for an end-year tax reaching $376 in North America and $548 on average for the rest of the OECD.[21] For a more limited target, such as stabilizing carbon emissions at the 1990 level, estimated carbon taxes are much smaller. Thus, the Congressional Budget Office (CBO), has estimated that a tax of $30 a ton of carbon ($4 a barrel of oil) would stabilize U.S. emissions at 1990 levels by the year 2000.[22]

The political questionability of a carbon tax is the more evident when its differential impact on the fossil fuels is considered. The Congressional Budget Office has calculated that a tax of $30 a ton of carbon would equal 55 percent of the delivered price of coal and about 10 percent of the prices of oil and natural gas (all in 1994 dollars). An IEA model's results for higher carbon tax rates show even greater disparities in the effects on the prices of the different fuels.[23] Unyielding resis-

20. Andrew Dean and Peter Hoeller, *Costs of Reducing CO₂ Emissions: Evidence from Six Global Models*, Economics Department Working Papers 122 (Paris: OECD, 1992), p. 8, table 1; and Richard A. Bradley, Edward C. Watts, and Edward R. Williams, eds., *Limiting Net Greenhouse Gas Emissions in the United States*, Report to the Congress of the United States, vol. I, Energy Technologies, DOE/PE-0101 (Washington: Department of Energy, September 1991), p. xvi.

21. E. Lakis Vouyoukas, International Energy Agency, *Carbon Taxes and CO₂ Emissions Targets: Results from the IEA Model*, Economics Department Working Papers 114 (Paris: OECD, 1992), p. 20, p. 19, table 5. For a 3 percent CO₂ reduction target, the model results are a $700 North American tax and a $1,222 tax for the rest of the OECD.

22. Congressional Budget Office, *Reducing the Deficit: Spending and Revenue Options* (Government Printing Office, February 1993), p. 404.

23. Vouyoukas, *Carbon Taxes and CO₂ Emissions Targets*, p. 19, table 5.

tance from coal, an industry characterized by high amounts of sunk capital and relatively immobile labor, is to be expected.

A carbon tax could be a source of large revenues for governments. For some of the tax rates discussed in the literature, the initial and possibly the more lasting contractionary effect on national output could be painful. In a 1990 paper, a CBO simulation found that a carbon tax beginning at $10 a ton in 1991 and rising to $100 in 2000 might keep U.S. GNP after 1995 a steady 2 percent below what would otherwise have been expected.[24] An OECD survey of modeling for ten countries found that with various assumed tax rates end-year GNP would in all cases be lower than the base case.[25] These consequences presumably could be mitigated if other taxes were to be reduced or monetary policy eased, but the lingering adjustments imposed on energy users and producers would still entail costs.

To these disabilities of a carbon tax must be added the disparate positions of the OECD and the rest of the world. Currently, the OECD member countries account for nearly one-half of the global emissions of carbon dioxide.[26] But developing country populations and their economies will grow faster in the decades ahead. So, therefore, will their contributions of carbon dioxide to the atmosphere. Business-as-usual policies in the third world and Eastern Europe will over time make fruitless any unilateral efforts by the OECD to stabilize much less to reduce the growth of global emissions.[27]

It follows that if stabilization or absolute reduction is the

---

24. Congressional Budget Office, *Carbon Charges as a Response to Global Warming: The Effects of Taxing Fossil Fuels* (Washington, August 1990), p. 35.

25. Peter Hoeller, Andrew Dean, and Masahiro Hayafuji, *New Issues, New Results: The OECD's Second Survey of the Macroeconomic Costs of Reducing CO_2 Emissions*, Economics Department Working Papers 123 (Paris: OECD, 1992), p. 21, table 6.

26. World Resources Institute in collaboration with United Nations Environment Programme and United Nations Development Programme, *World Resources 1992–93*, pp. 346–47, table 24.1.

27. Bradley, Watts, and Williams, eds., *Limiting Net Greenhouse Gas Emissions*, p. x.

target, a worldwide cooperative effort well beyond that proposed by the climate change treaty agreed at the Rio conference will be necessary. Yet ideas such as a universal carbon tax or a scheme for international trading in emission permits are hopeless nonstarters for an indefinite period ahead.[28] (The non-OECD countries will not levy a tax that will impair their growth prospects and the OECD countries will see that tradeable permits imply huge increases in flows of resources to developing countries and perhaps to Eastern Europe.) Until something now unforeseeable happens to make emission controls much cheaper or to convince the world that the potential for catastrophic climate change is more certain and more imminent than is now the case, worries about greenhouse gases will most likely induce only modest and tentative countermeasures in the rich countries and even more limited actions in the poor.

Environmentalism, nonetheless, seems bound to help restrain the growth of oil demand in the period ahead. Even cost-inefficient controls on oil use and subsidies for oil substitutes will probably bring net reductions in demand. If environmentalist support serves to make higher oil consumption taxes feasible, the effect almost surely will be greater. The Clinton administration's proposed energy tax, though designed primarily for fiscal objectives, would have reduced oil consumption by an estimated 400,000 barrels per day by the year 2000.[29] If actions by the United States and other OECD nations were to be matched by reforms of the economically absurd pricing and subsidy practices prevalent in the developing world and Eastern Europe the effect on oil demand could be more consequential still. As noted, the green movement is at a relatively early stage outside the main industrial countries, but together

28. For discussion, see Thomas C. Schelling, "Some Economics of Global Warming," *American Economic Review*, vol. 82 (March 1992), pp. 11–14.

29. Because the energy tax, as drafted, would bear more heavily on oil than on coal, its contribution to the reduction of carbon dioxide emissions would be smaller than that of a true carbon tax. Overall, however, it would advance environmentalist goals and so should have the backing of environmental groups.

with other forces for change it could in due course have a noticeable depressive effect on oil consumption, where that seems bound to increase most rapidly. Summed up, however, the potential damage to the environment would have to look more certain and its dimensions more forbidding before ecological concerns in themselves would bring about a large reduction in future world oil consumption.

CHAPTER FOUR

Political Uncertainties

PAST OIL SHOCKS have had their proximate origins in local war, political upheavals, and a grab for the control of Gulf oil, all of these of course in the Middle East. Whether similar events will recur, and have similar consequences for the price of oil, is neither assured nor foreclosed. The current scene offers important grounds for optimism. To project such optimism through the 1990s and the next decade could, however, prove to be expensively wrong.

Two recent, unexpected developments have much improved the prospects for a regional setting favorable to dependable oil supplies. One was the Gulf War, the other the collapse of the Soviet Union.

Iraq's defeat has removed for an indefinite period the threat presented by the most overtly aggrandizement-minded of the Middle East states. Much is being made of the survival of Saddam Hussein's army and his shows of defiance to the United Nations' inspectorate. In fact, his forces suffered heavy losses of men and equipment and at the war's end he was forced to accept the terms of a humiliating peace. He has not been able to sell his oil since mid-1990. The embargo on exports to Iraq may be leaky locally, but it has remained effective insofar as access to the principal suppliers of industrial and military goods is concerned.

To rebuild Iraq as a leading regional power will require that markets for Iraqi oil be reopened and that foreign banks and businesses be willing once more to offer credits to this already heavily indebted country. So long as Hussein's presence continues to make Iraq an international pariah, these things are not likely to happen. And even after his departure, recovery cannot be immediate. Nor will a return to his aggressive posture be appealing to a successor regime—possibly the inheritor of a truncated Iraq—at any early date.

The Gulf War has also served as a warning signal to Iran. With Iraq incapacitated, Iran is by default the dominant Persian Gulf nation. Along with ambitions for extending its role well beyond the Gulf, Teheran doubtlessly would like to exercise a decisive influence on the price of oil, that is, on the policy of Saudi Arabia. Any temptation to pursue that objective by force or the threat of force must take account, at least for some time to come, of the formidable international response aroused by Iraq's 1990 venture.

When the Gulf crisis erupted, the Soviet Union had already moved away from its long-time posture as the superpower rival to the United States in the Middle East. Subsequent developments have effectively removed it as a political and military force in the region. Consequently, the Arab-Israeli confrontation has been altered in an important respect. In particular, with Iraq subdued and Syria having lost its Soviet patron, the core of Arab resistance to serious consideration of any advance toward peace with Israel has been much eroded.

Eventually, the Russian Republic will be able again to claim to have a voice in Middle Eastern affairs. That can come, however, only after the transition from the failed communist experiment has made definitive progress toward a workable system and polity, not only in the Republic but also in its relations with the other new states deriving from the erstwhile Soviet Union. Meanwhile, and this interval promises to be an extended time, America will be unchallenged as the leading foreign quantity in the Middle East. To the extent that any external force can help to shape regional events, it will be the United States.

Early signs have been favorable. The Soviet Union's demise, an Israeli election, and patient American diplomacy have improved the chances for progress toward something better than the shaky armistice that has obtained between Israel and its neighbors, Egypt excepted, since the 1973 war. No doubt, as Henry Kissinger has argued, a comprehensive agreement, encompassing peace treaties, final boundaries, and the status of Jerusalem, is beyond present reach.[1] Nevertheless, the principal parties have reasons for pursuing, or at least not blocking, interim arrangements that will give hope that a new Arab-Israeli war can be pushed further away from being a possible cause for another oil crisis.

For Israel the alternative is permanent tension and the likelihood of increased isolation. For the governments of Egypt and Jordan ongoing failure to find relief for Palestinian grievances can only heighten the danger of runaway Islamist extremism at home. For Syria a state of undiluted confrontation with Israel gives no prospect for recovery of the territory lost in 1967. For Saudi Arabia, moves contributing to regional stability are bound at bottom to be welcome. Iran alone may stand in opposition to peace negotiations but, except for the nurturing of terrorists, will have little power to interfere in what are, after all, Arab affairs.

Cause for qualified optimism thus seems justified. The defanging of Iraq, the elimination of the Soviet Union as a regional force, and the improved outlook for Arab-Israeli relations together argue for a time of relative calm—at least in respect to the risk of sudden interruption of the flow of oil from the Middle East. Whether this situation can be a durable one remains to be seen.

Political change is certain, as it is everywhere, but in the Middle East the possibilities could be especially disturbing.

Egypt has been a rock of moderation since the Camp David accords, a virtual guarantee against another Arab-Israeli war. As is now true in Algeria, Tunisia, the Sudan, and Jordan, Islam has increasingly become a political as well as religious force in Egypt. There the main Islamic movement,

1. Henry Kissinger, "The Path to Peaceful Coexistence in the Middle East," *Washington Post*, August 2, 1992, p. C7.

the Muslim Brotherhood, operates as an (officially unrecognized) political party with an essentially moderate platform. It leads the parliamentary opposition and has placed itself apart from Egypt's extremist Islamist elements.

But Egypt is a very poor country with large restless groups in its urban populations and in its south. President Hosni Mubarak, aided by American military and economic grants and by the flow of remittances from the Gulf states, has held extremism in check for more than a decade. His successor may need to be equally skilled in dealing with an always potentially volatile situation. If the successor were to fall short, a main pillar of Middle East stability could be shaken and the odds against a renewed Arab-Israeli war reduced.

The most populous (and growing at 3.5 percent in 1992) nation in the Middle East is Iran. It is emerging from an era of theocracy, marked by fiercely anti-Western public and official attitudes and by a lengthy, debilitating war provoked by Iraq. Now, with that war well over and the more disruptive and costly aspects of the Khomeini period modified, the Iranian economy can be expected to make more effective use of the country's sizable pools of resources and skills. By the standards prevailing in the region, the political system that has come into being de facto since Khomeini represents an advance toward a more open system. Eventually, the attractions of support for terrorism may disappear, and normal relations with the United States and other Western countries may come to exert a chastening influence on Iran.

At the same time, Iraq's weakness and possible dismemberment, the breakup of the Soviet Union, and the upsurge of Islamist political forces in North Africa and elsewhere can hardly fail to stimulate ideas of extending Iran's influence into Central Asia and into the Arab world beyond the Persian Gulf. Conceivably, attempts to realize these prospects could so unsettle intraregional stability as to endanger the flow abroad of Gulf oil. Already the small Gulf states, observing Iran's arms purchases (including submarines from Russia), nuclear program, and seizure of a disputed island in the Gulf, see a new regional threat replacing Iraq. Or, if and as the United States tires of the burden of keeping a ready force in the Gulf region, a well-armed but economically hard-pressed Iran might choose to emulate Saddam Hussein and seek to use military pressure to exert control over Saudi Arabian policy, possibly leading to a sharp and serious reduction in Saudi exports.

Political developments in Saudi Arabia could present a threat to the flow of oil. The familial autocracy that has governed the country within its present borders since, more or less, the end of World War I is an anachronism—a hereditary house that has managed to keep for itself near-absolute power in a world where this form of rule has all but disappeared. The regime has not gone unchallenged. Strikes in the oil fields, sporadic protests by the Shiite minority, and opposition from purist Islamic elements have occasionally and briefly seemed to threaten the nation's political tranquility. But neither these events nor the inevitable imprint of modernization on an erstwhile nomadic society, nor the exposure of tens of thousands of young Saudi students to Western ideas and ways of life, seem to have shaken the regime's control. By all accounts, the armed forces remain loyal and the ulema, the senior Islamic clergy, dependent unlike their counterparts in Iran on governmental financial support, have been prepared, if sometimes reluctantly, to confer on the royal house the status of religious legitimacy.

In the face of all these conditions, the kinds of internal events that might put Saudi oil at risk can be no more than grossly speculative. Still, a seventy-year-old absolute monarchy, even a rich and inherently cautious one, cannot be considered everlasting. Perhaps change will come in orderly ways. King Fahd has promised a consultative council, appointive, which could evolve into a form of legislature and open the road to an eventual constitutional monarchy. Or a less benign pattern of change—for example, disputes over the line of succession in the royal family—could emerge without affecting the flow of Saudi oil to the rest of the world.

But change could be disorderly and violent, as, say, in the

form of a coup by disaffected elements of the military. Again, if change seemed to point to the development of a more secular Saudi Arabia, a political-religious upheaval on the Iranian model could be envisioned. Domestic turmoil could present a dilemma to the house of Saud's ally and protector, the United States, for intervention in an internal dispute might well heighten the threat to oil exports.

CHAPTER FIVE
Possibilities for Oil Disruption

WE BELIEVE THAT over the next twenty years the world economy will become more dependent on oil from the Persian Gulf than ever before. We also believe that political developments make the region seem less dangerous than at any time in the past four decades, even offering the promise of progress in resolving or easing some conflicts. Nonetheless, oil and the deepseated roots of political instability in the Middle East are a dangerous mixture. Twenty years is a long time in which to offer assurances about the unpredictable. For the next twenty years quasi-military threats, coups, or war cannot be ruled out and for that reason neither can oil supply interruptions leading to another price shock.

Although the probability of such interruptions has diminished, their cost, should they come about, will still be high. How much oil could be affected by a crisis at different points in the future? What might be the consequences for oil prices, assuming for present purposes the absence of defensive measures by the OECD countries? It is almost axiomatic that a future crisis in the region will not be a replay of the past. Nonetheless, the following stylized scenarios, including an examination of past crises in a new setting, can help to place the risks in useful perspective.

—*A selective political embargo.* An Arab political embargo targeted against the United States or any combination of OECD countries, whatever the motivation, would have little effect in any foreseeable circumstances. Enough oil would be available from other sources to meet the needs of the targeted

62

countries, and the international industry would have the flexibility to reallocate oil whether Arab destination controls are imposed or not.

—*Across-the-board cuts in exports.* A more serious matter would be across-the-board percentage cuts in exports by most or all of the Middle East producers in reaction to political or military developments in the region. The first oil shock was touched off by just that kind of action, taken without Iraq and with Libya joining only belatedly. A replay of the 1973 scenario would, however, require that Saudi Arabia again be a participant. In the light of subsequent events up to and through Desert Storm, Saudi economic and security interests, and those of Kuwait and the United Arab Emirates, would argue strongly against participation. Without the support of these three countries, a new embargo would cause no more than a temporary stir in oil markets and would soon prove self-defeating. A change in Riyadh to a regime with an overriding anti-Western orientation would be necessary to alter that judgment.

—*A repeat invasion of Kuwait.* In view of Iraq's continued espousal of irredentist claims on Kuwait, the consequences of another invasion a decade from now are worth examining, not because of its plausibility but as a test of the flexibility of the oil market. Early in the next century, Iraq and Kuwait could be producing a total of 6–7 mmb/d compared with 3.5 mmb/d just before Iraq moved into Kuwait in August 1990. At that time just about enough excess capacity was available to offset the shortfall when the Western blockade cut off exports from the two countries. Commercial stocks, moreover, were unusually high. Nonetheless, oil prices almost doubled in two months. Our forecasts have assumed that a more or less normal cushion of 3 to 4 mmb/d would be maintained by OPEC countries, mainly by those in the Persian Gulf, through the projected period. If so, the removal of Iraq-Kuwait production from the market could not be made up by other exporters. The physical shortage of some 3 mmb/d would cause a much larger flare-up of prices than in 1990, the amount and duration depending also on the market's view of how long it would take for production to recover and normal exports to resume.

—*Military conflicts.* There are more extreme possibilities. Suppose that by the year 2000 military action involving Iran and one or more Arab countries broke out or that another Arab-Israeli war had begun. Such conflicts could make the Persian Gulf an active war zone, threatening the 22 to 24 mmb/d expected to be coming from the area with war damage or sabotage. Pipelines, which might be carrying nearly half this amount, would be vulnerable to air attacks on pumping stations and tank farms, as allied damage to Iraqi installations demonstrated in Desert Storm. The experience of the Iran-Iraq war suggests that tanker traffic would be less liable to disruption, but measures would still be needed to make sure that shipowners and crews would be willing to ply the Gulf. In the end, determined Western action to protect pipelines and sea lanes would be needed to avoid large-scale disruption of oil supplies. If that took time to organize and carry out, as it likely would, a temporary loss of at least one-fourth of the flow of oil from the Persian Gulf, or 5 to 6 mmb/d, can readily be envisaged. By the year 2010 when 28 to 30 mmb/d could be coming from the area, the size of an oil interruption arising from a military conflict could be larger.

—*Trouble in Saudi Arabia.* Another extreme case, postulated as taking place a decade from now, would involve an interruption of oil production in Saudi Arabia. The causes might be civil unrest and strikes in the oil fields leading, as in Iran, to revolution or to a military coup. Or external pressure from militant governments in the area might lead to attacks on oil installations. None of these contingencies can easily be foreseen in present circumstances, but in view of Saudi Arabia's crucial role in the world energy system, their consequences need to be considered. During the first decade of the next century, the world market probably will be relying on Saudi Arabia for 10 to 15 mmb/d, compared with about 8 mmb/d today. The Saudis recently increased pipeline capacity from 3.2 to about 5 mmb/d and may decide on a further expansion as their production grows. And a large share of OPEC's spare capacity over the future, easily half, is likely to be in Saudi Arabia. Thus a curtailment of Saudi exports because of physical damage to its production or transportation facilities or because of political decisions emanating from turmoil and change would be bound to cause an upheaval in oil markets. It could mean a very sizable, temporary reduction in world oil supply and with it the removal of much of the spare capacity that had been relied on to cushion such a loss.

In short, without assurance of a Middle East free of serious conflicts or political upheaval, another oil supply interruption of sizable proportions is certainly possible. In 1973–74, the Arab oil embargo caused an average net shortfall of 2.5 mmb/d over a six-month period. In 1978–79, the Iranian revolution caused a net shortfall of about 2 mmb/d for four months. Each time the economic consequences contributed importantly to a world recession. Prudent policy has to take into account the possibility of another such occurrence.

CHAPTER SIX
Policy Responses

CIRCUMSTANCES SURROUNDING ANY future interruption in the supply of oil will determine the damage it can cause. These circumstances include the strength or weakness of the world economy at the time, the ability to use macroeconomic policy as a defensive measure, and the effectiveness of the system established by the members of the International Energy Agency to manage an oil disruption. No doubt these factors will differ from those that applied in the past. We will start, however, by reviewing past events and then ask what might be done to strengthen defenses against a future shock.

History

Three oil shocks in two decades are testimony to the volatility of the oil-producing Middle East region. Each arose from different causes, produced different reactions, and varied in its economic impact.

1973–74

No event of the period following World War II had so sharp and pervasive an impact on the world economy as the series of price and supply shocks to the oil market that followed the outbreak of the Arab-Israeli war on October 6, 1973. For more than a year uncertainty grew, and pessimism seemed to feed on itself. Indeed the economic, political, and psychological repercussions of what became known as the "energy crisis" caused widespread questioning of the capacity of the world economy to adjust to the new situation at tolerable cost.

The sequence of events started when the Arab oil-exporting countries in October 1973 announced they would cut exports by 10 percent, and threatened subsequent cuts of 5 percent a month until their demand for the evacuation of Israeli forces from Arab territories was met. Subsequently, they decided to embargo shipments of their oil to the United States, South Africa, the Netherlands, and Portugal. As it turned out, the cuts were less severe and did not last as long as had been anticipated. The ability of the oil companies to reroute oil from other exporters to the embargoed countries made the selective destination controls ineffective although they added to uncertainty and tension in the market. For all practical purposes, use of the oil weapon ended by April 1974. Between November 1973 and April 1974, the net reduction in exports averaged 2.5 mmb/d, or 5.5 percent of consumption in oil-importing countries, even with all other exporters producing at capacity.

Spot prices, however, began to escalate immediately after the Arab exporters' announcement of production cutbacks. Encouraged by that response, OPEC at meetings in October and December in effect quadrupled the official or contract prices at which most oil moved. With usable emergency reserves almost negligible, confusion in the market, and widespread uncertainty about the future supply of oil, a competition to develop preferred bilateral relations with exporters and attempts to build stocks ensued. The price run-up was fully predictable.

The increase in crude oil prices meant first of all that oil-exporting countries in 1974 gained an additional $75 billion a year in export receipts, which for the oil-importing countries meant a terms-of-trade loss equal to 1.5 to 2.0 percent of their GDP.[1] In the OECD countries, moreover, a transfer of income

1. Edward R. Fried and Charles L. Schultze, "Overview," in Fried and

occurred between domestic producers and consumers of energy, estimated at 1 percent of GDP.[2] So large a transfer of income could not be spent quickly enough to avoid a drag on aggregate demand in the oil-importing economies. Furthermore, the increase in oil prices added about 2 percentage points to the general rate of inflation, which with the increase in prices of other fuels and the wage-price spiral they encouraged, increased the direct inflationary effect several times over the next few years.[3] This occurred, moreover, when the synchronized world economic expansion of 1972–73 had already put pressure on industrial capacity and caused a boom in prices of industrial materials. A worldwide drought in the 1972–73 crop year compounded these commodity price pressures by causing a tripling of grain prices. Finally, the huge oil-induced changes in international accounts put pressure on the operation of the international financial and trading system.

These economic problems and the economic policies adopted to cope with them combined to reduce output in the OECD countries below what otherwise might have been expected. Quantitative estimates of how much damage the oil shock itself caused are speculative at best, partly because other strong adverse forces were already leading to some weakening before the oil shock hit. Studies at the time suggest that economic output in OECD countries in 1974 was 2 to 3 percent below the preembargo outlook because of the impact of higher oil prices.[4]

For the oil-importing developing countries, the additional cost of oil amounted to $10 billion in 1974, somewhat more than their net inflow of concessional capital. Their export receipts, furthermore, suffered from the oil-induced contraction

Schultze, eds., *Higher Oil Prices and the World Economy: The Adjustment Problem* (Brookings, 1975), p. 13, table 1-3; and authors' estimates.

2. John Llewellyn, "Resource Prices and Macroeconomic Policies: Lessons from Two Oil Price Shocks," Working Papers 5 (Paris: Organization for Economic Cooperation and Development, April 1983), p. 5.

3. Llewellyn, "Resource Prices," p. 5.

4. Fried and Schultze, "Overview," pp. 18–27. See also "The Impact of Oil on the World Economy," *OECD Economic Outlook*, no. 27 (July 1980), pp. 114–30.

of growth in the industrial countries, and they suffered terms-of-trade losses as well. Nevertheless, they managed as a group to escape the recession, principally because they borrowed on a large scale from foreign banks, a process welcomed at the time as a useful recycling of OPEC financial surpluses. Economic growth held at about two-thirds the preshock level, but the borrowing that helped make growth possible contributed to the debt problem that depressed some of these economies in the 1980s.

Policy responses in the industrial countries eventually proved accommodative. Most central banks were tightening monetary policy earlier in the year in response to overheating. In the United States and Germany monetary policy began to relax in the aftermath of the oil shock. Japan, notably, had been tightening early in 1973 in response to the emergence of double-digit inflation and continued to do so after oil prices hit; nonetheless the earlier buildup of demand pressures was so strong that inflation rose above 20 percent early in 1974. In that year, for the first time in the postwar period, Japan experienced negative economic growth. In all of the major OECD countries real interest rates declined, turning negative everywhere in 1974 except in Germany.[5]

As the deflationary consequences of the oil shock became apparent, most governments shifted toward fiscal expansion. Some of this shift happened automatically because of the operation of fiscal stabilizers when the recession took hold. Germany expanded in 1974; the other major countries in 1975. For the Group of Seven, according to OECD estimates, "the discretionary shift of fiscal policy towards expansion in 1974 and 1975 together was very nearly 2 per cent of their combined GNP."[6]

In considering these policy responses, two points deserve

---

5. This discussion of monetary policy reactions draws heavily on Michael M. Hutchison, "Aggregate Demand, Uncertainty and Oil Prices: The 1990 Oil Shock in Comparative Perspective," BIS Economic Papers 31 (Basel, Switzerland: Bank for International Settlements, August 1991), pp. 39–43.

6. Llewellyn, "Resource Prices," p. 6.

special mention. As John Llewellyn points out, it was not readily apparent when the first oil shock hit that the consequences
could simultaneously be inflationary and deflationary (reducing real demand).[7] Second, the oil shock helped imbed inflation more firmly into the system. For the OECD as a whole,
inflation reached a peak of about 13 percent in 1974 and,
although declining thereafter, was still as high as 8 percent in
1978, twice the level of the 1960s. These aspects of the 1973–
74 experience influenced government reactions to the next oil
shock.

1979–80

Unrest, turmoil, and then revolution in Iran brought on
the second oil shock in 1979–80. It started with strikes in the
oil fields, reducing production from 6 mmb/d to an average of
4 mmb/d in the fourth quarter of 1978. Increased production
from other exporters, chiefly Saudi Arabia, quickly made up
most of that shortfall. In addition, commercial stocks initially
fell in response to normal seasonal needs and the modest net
shortage in supply. Spot prices rose only moderately on the
market's expectation of slack when Iranian production re-
turned to normal.

Instead, the shah fell in January 1979. As the secular and
clerical factions fought for control of the revolution, oil production came to a near halt. During the first quarter of 1979,
the shortfall in Iranian production averaged 4.2 mmb/d. Increased production by other exporters made up half this
amount, leaving a deficit in supply of 2 mmb/d, or 4.3 percent
of consumption in the oil-importing countries. On top of this,
it was feared that Ayatollah Khomeini would put a permanent
lid on Iranian oil production to promote austerity and beyond
that would try to spread his doctrine to other Muslim countries,
including the Persian Gulf oil exporters.

7. Llewellyn, "Resource Prices," pp. 4–9.

Spot prices then began to rise in earnest. In response, the oil companies, sometimes prodded by governments, scrambled to build stocks in anticipation of further price increases, a self-fulfilling strategy. OPEC then used the escalation of spot prices as an excuse to raise official prices, thus continuing the upward spiral, with Saudi Arabia alone attempting, unsuccessfully, to apply the brakes.

This sequence resulted in oil prices increasing 150 percent during 1979 and 1980, with little justification from the size of the oil supply shortfall. At the beginning of 1979, the IEA forecast that the shortfall for the year would be 5 percent of consumption, not enough to trigger the organization's collective emergency defense program. By April, if measured by current consumption requirements, no shortage existed. Stockbuilding, taking place in a fragile market, was responsible for the oil supply deficit and the rising prices.[8] During the year primary stocks increased by 1.5 mmb/d and a one-time surge in secondary and tertiary stocks (accumulated by end product users) accounted for an estimated increase in demand of 2 mmb/d for a three-month period.

The jump in oil prices transferred $170 billion from net oil-importing countries in the OECD area to oil exporters, a transfer of about 2 percent of GDP. It occurred before the effects of the first shock had been fully absorbed and led to similar consequences—a drag on aggregate demand, inflationary pressures, and a distortion of international payment positions. Calculations by the OECD secretariat suggest that these effects caused OECD output in 1980 and 1981 combined to be about 3 percent below what it otherwise would have been. About half this loss is attributed to the reduction in aggregate demand induced by oil prices and the other half to the OECD

8. In the United States, price controls although being gradually removed beginning in 1978, continued to subsidize oil imports and encourage the buildup of stocks. Refiners at the time could still average the cost of price-controlled domestic oil with higher-priced foreign oil, thereby weakening the restraining effect of rising world oil prices on stockbuilding.

policies put into effect to counter the oil-induced inflationary pressures.[9]

This time the oil-importing developing countries came out much worse. The jump in oil prices cost them an estimated $25 billion to $30 billion in additional import costs, which they could not finance by new borrowing from commercial banks because their debt overhang was already too high. They soon faced a liquidity, then a solvency, crisis. With some notable exceptions in East Asia, economic growth fell sharply, becoming negative in per capita terms, and has taken a long time to recover.

In contrast to 1973–74, OECD governments responded to the second oil shock by seeking to combat its inflationary effects rather than to support demand in the short run. A growth expansion had been under way in 1978, but inflationary forces varied among the Group of Seven (G-7) countries. Germany and Japan, focusing on the money supply since the mid-1970s, had brought prices under control and were intent on not accommodating new external shocks. Among the other countries, inflation was still a disturbing concern; earlier in 1978, the United States and the United Kingdom had started to tighten monetary policy. As oil prices rose, so did interest rates. In October 1979, concerned over prices that were already high and going higher because of oil, the Federal Reserve began to target the money supply in earnest. Short-term interest rates in the United States went to high double-digit levels, putting pressure on monetary policy and interest rates in other countries. In response to the prevalence of monetary restriction, consumer prices in the G-7 countries fell from a peak of 10.5 percent in 1980 to 6.7 percent in 1982, lower than the average before the second wave of increases in oil prices began.

9. Llewellyn, "Resource Prices," p. 15, table 6. These results are a downward revision of earlier OECD estimates, principally because OPEC spent its additional revenues faster than had been anticipated. See, for example, Sylvia Ostry, John Llewellyn, and Lee Samuelson, "The Cost of OPEC II," *OECD Observer*, no. 115 (March 1982), p. 37–39; and *OECD Economic Outlook*, no. 31 (July 1982).

Positive real interest rates became the norm rather than the exception.[10]

Concern over budget deficits before the second oil shock restricted the possibilities of using fiscal stimulus to counter the oil drag on aggregate demand. Hutchison characterizes discretionary fiscal policy in the G-7 countries in 1980–81 as "generally neutral or slightly contractionary," again in contrast to 1973–74.[11] Only Japan applied fiscal stimulus during this period, which along with a strong performance in containing inflation, enabled it this time to go through an oil shock with respectable economic growth. The United States moved strongly to fiscal expansion with a large tax cut in 1981, mainly however for supply-side rather than antirecessionary reasons.

1990

Iraq's invasion of Kuwait on August 2, 1990, followed quickly by the United Nations blockade of the two countries, meant an initial loss of 4.3 mmb/d of crude oil, products, and other liquids to the world market. About enough spare capacity was immediately available or mothballed in other OPEC countries, chiefly Saudi Arabia, to make up the loss. Within a month these countries had geared up to fill the gap. By October, OPEC production was less than 1 mmb/d short of the July level; by December, the remaining difference had been eliminated. Furthermore, OECD commercial stocks just before the invasion were abnormally high, as were the floating stocks held abroad by Saudi Arabia, Iran, and Venezuela.

Despite this reasonably sanguine outlook for the oil supply, average monthly prices jumped from almost $16 a barrel in July to about $34 in September. The primary reasons apparently were uncertainty at the start about how much OPEC

10. Hutchison, "Aggregate Demand," describes monetary reactions to the second oil shock in pp. 43–46.
11. Hutchison, "Aggregate Demand," p. 48.

spare capacity was available and would be put to use; and, subsequently, fears about the destruction of oil installations elsewhere in the Gulf should war break out between Iraq and the allied forces. In response to the resulting speculative and precautionary demand, commercial stocks increased during the fourth quarter, counterseasonally, and secondary and tertiary stockbuilding again became evident. Prices fell moderately in December, strengthened just before war broke out, and collapsed when the military outcome became apparent. Only on the outbreak of the war, in a move calculated to preempt panic, did the IEA finally announce that some government stocks would be released to the market.

In all, average oil prices between August and December 1990 were about $10 a barrel higher than had been expected before the invasion. On that basis, the IMF estimated that in 1990 this jump in oil prices raised consumer prices in the industrial countries by 0.5 percent and short-term interest rates by 0.25 percent. These consequences together with a deterioration in the terms of trade and the external balance and a decline in real disposable income are estimated to have reduced real GDP in the industrial countries by about 0.25 percent in 1990, or by about $25 billion, with a gradual return to trend growth expected during 1991 and 1992. In the net debtor, oil-importing developing countries, real GDP is estimated to have been lowered by 0.6 percent in 1990 and 0.4 percent in 1991, including the effects of the losses of workers' remittances from the Middle East.[12]

Monetary and fiscal policies generally reflected a continuation of preinvasion positions rather than a response to higher oil prices. The United States continued to ease monetary policy in reaction to the recession as did France. In contrast, Japan, Germany, and the United Kingdom continued to tighten. Budget balances deteriorated mainly as the result of

---

12. International Monetary Fund, "Oil Price Assumptions and Economic Effects of Crisis in the Middle East," *World Economic Outlook, May 1991* (Washington, 1991), pp. 21–25.

the slowdown in economic activity, not because of discretionary fiscal stimulus.[13]

How Much Economic Damage Was Avoidable?

In essence, the unusually large damage from these episodes occurs because modern industrial economies are not sufficiently flexible to adjust smoothly to large external shocks. Other prices do not fall when oil and other energy prices rise, and wages tend to be rigid as labor tries to avoid the real income losses from increased energy costs. The sudden huge income transfers between energy producers and energy consumers within and between countries cannot be managed so as to avoid depressing aggregate demand without adding to inflationary pressures. Distortions in international payment balances constrain trade and monetary policies. All these conditions add to uncertainty and weaken business confidence, which dampens investment and consumption. The feedback effects of similar developments in other oil-importing countries add to the downside pressures.

Successful policy responses and a learning experience can help. In general, the emphasis against accommodation probably reduced the damage in the second shock. So did the fact that wages were more flexible. As the most notable example of improved policy management, Japan succeeded in containing its cost structure and adjusting rapidly the second time around compared with its exceptionally poor performance the first time. The third shock did not last long enough to take the measure of OECD fiscal and monetary policy. The IEA decision not to release stocks earlier, however, is certainly questionable; an early, coordinated release of emergency stocks would have done much to inspire confidence and temper the price increase that took place.

Nonetheless, large disruption costs over and above terms-

13. Hutchison, "Aggregate Demand," pp. 46–52.

of-trade losses probably were unavoidable. The complexities involved—the inevitable uncertainty about the duration and ultimate size of the interruption in oil supply, the internal political constraints on action, and the difficulties of coordinating policies among the major countries given the likely differences in economic conditions among them—limit the room for maneuver. As Charles Schultze recently wrote, "Because it simultaneously sets in motion both recessionary and inflationary forces, a sudden disruption of oil supplies accompanied by a large run-up in oil prices is virtually impossible for economic policymakers to handle. Measures that ease the recessionary problem exacerbate the inflation problem, and vice versa."[14]

If foreign or military policy cannot prevent the jump in oil prices in the first place, reducing its extent should be the central object of policy. Such action comes down to enacting emergency nonprice measures to reduce the demand for oil in the short run; avoiding a competitive scramble for stocks, which in each of the three shocks exacerbated the effects of the supply interruption; and adding to supply through the timely release of stocks. To be fully effective, action must be multilateral since oil prices are related to what happens on the world market; whatever defensive action any one country takes is made more, or less, effective depending on what other countries do.

Coordinated Action: The Emergency Program of the International Energy Agency

To explore the possibilities for mitigating the economic costs of an interruption in the oil supply, we hypothesize a crisis at some future time, say, around the turn of the century. What economic damage might be caused, assuming no defensive actions are taken? And how much of this damage might

14. Charles L. Schultze, *Memos to the President: A Guide through Macroeconomics for the Busy Policymaker* (Brookings, 1992), p. 139.

be avoided by emergency measures taken by the member countries of the International Energy Agency (IEA)?

A Simulated Interruption

Assume that war or other turmoil in the Middle East around the year 2000 cut off 6 mmb/d of oil exports from the affected countries. Assume also that 2 mmb/d could be made up from the use of spare capacity in other OPEC countries. Thus a net OPEC export shortfall of 4 mmb/d would occur, and the authorities would have no reliable way of knowing at the outset how long the shortfall would last.

That much of an oil shortfall at that time would constitute 5.5 percent of world consumption and 7.5 percent of consumption in oil-importing countries. In response, oil prices might double, rising, say, from $25 to $50 a barrel in 1990 dollars.[15] This estimate would be conservative based on past performance. For the OECD countries, this price jump would impose a terms-of-trade loss, or income transfer, of almost $400 billion, or 1.8 percent of their projected income.

The OECD has estimated that a price increase of $10 a barrel in September 1990 could reduce GDP in OECD countries by an average of 1 percent a year in 1991 and 1992, the amount varying slightly depending on whether the response emphasized tightening monetary or fiscal policy. The results would be disproportionately worse for much higher price increases because of the deterioration in private confidence and investment.[16] Economic conditions a decade from now, however, might permit greater policy flexibility and therefore somewhat lower losses. Oil costs in relation to GDP, although much lower than during the first two shocks, should be only moder-

15. Prices and consumption in OPEC countries and in non-OPEC developing oil-exporting countries are not likely to change. The burden of adjustment would then fall entirely on the OECD countries and the oil-importing developing countries. The short-run price elasticity is assumed to be -0.1 and market panic and stockbuilding to be moderate.

16. *OECD Economic Outlook*, no. 48 (December 1990), pp. 38–41.

ately smaller than they are now. For present purposes, therefore, an increase of $25 a barrel in the price of oil a decade from now is taken to reduce OECD GDP by 2–2.5 percent a year for two years, or by roughly $400 billion to $500 billion a year. For the oil-importing developing countries, the OECD secretariat estimates that the economic costs are likely to be severe because of the continuing high oil and energy intensities of their economies.[17]

Speculative as they are, these numbers show that another large oil price shock could again push growing economies close to or into recession. Can the costs be reduced?

Defensive Measures

The OECD countries established the International Energy Agency in 1974–75 in response to the political and economic consequences of the Arab use of the oil weapon.[18] Its formal emergency response system, designed with that event in mind, goes into effect when an interruption causes a loss of 7 percent of supply, the size of the loss from the embargo. Without detailing the complexities, IEA mechanisms would deal with a supply interruption by the following:

—Offsetting part or all of the shortfall by having member countries reduce demand through nonprice measures (for example, speed limits, driving restraints, fuel switching) or by drawing down stocks, or both, on the basis of an agreed-on formula; and

—Allocating oil among members essentially by redirecting imports to share the shortage, again on the basis of an agreed-on formula.[19]

---

17. *OECD Economic Outlook*, no. 48, pp. 41–43.

18. France participated in the discussions in 1974 leading to the formation of the IEA but decided not to join. It subsequently cooperated with IEA policies and became a member in 1992.

19. Daniel B. Badger, Jr., "International Cooperation during Oil Supply Disruptions: The Role of the International Energy Agency," in George Horwich and David Leo Weimer, eds., *Responding to International Oil Crises* (Washington: American Enterprise Institute for Public Policy Research, 1988), pp. 1–16.

The objective of these actions is to dampen the predictable surge in the price of oil. Nonprice measures, although entailing other economic costs, would help to contain price increases. Releases from stocks would also reduce the pressure on price. And, it is argued, oil sharing would moderate speculative pressures by reducing the need for competitive bidding on spot markets. To make certain that stocks could be drawn upon, member governments agreed to build up stocks equal to ninety days of imports; this obligation could be met by government-held stocks or by requiring the private oil industry to add to its stocks.

As a practical matter, the IEA could have done little to lessen the damage from the second oil shock. The shortfall, technically, was not large enough to trigger the emergency system even though the huge price increase that occurred did indeed create an economic emergency. Stocks were not yet large enough to be of use, especially in the atmosphere of panic that prevailed, even if informal coordinated drawdowns could have been arranged. Nor were IEA exhortations effective, at least initially, in restraining the scramble for commercial stocks.

It soon became evident that the IEA formal response system was defective because it did not require action until the shortfall reached 7 percent. Yet a smaller loss demonstrably could cause extensive damage. Critics also contended that the oil-sharing system was cumbersome and, if ever put into operation, would interfere with efficient market adjustment.[20] Meanwhile, the United States, Japan, and Germany, mainly, had been building strategic, or government-held, stocks that could be used unambiguously in an emergency. Stocks were now seen to be the most effective feature of the emergency response system.

In July 1984, the IEA, by ministerial agreement rather than amendment of the treaty establishing the organization, added an informal response plan to the system. Stock draw-

20. See George Horwich and David Leo Weimer, eds., *Responding to International Oil Crises* (Washington: American Enterprise Institute for Public Policy Research, 1988), for essays by proponents and critics of the IEA oil-sharing system.

downs could be brought into play in any significant disruption and were expected to be the first, not the last, line of defense. Stocks held by companies, equally with government stocks, could be drawn upon when allowed and ordered by governments. Demand restraint measures, when governments preferred this route, were given equivalent credit to stock drawdowns. Thus, even without formal obligations, it became possible through joint consultation in an emergency to work out a reasonable sharing of responsibilities.[21]

This informal system came into use in the aftermath of the invasion of Kuwait. When the allied forces made their countermove, the IEA announced on January 17, 1991, that 2.5 mmb/d would be made available (2 mmb/d from stock drawdowns and the rest from demand restraint) and urged companies and consumers to restrain purchases. As noted earlier, this announcement combined with the quick allied military success immediately caused prices to fall to precrisis levels.

What is puzzling about the IEA performance during the third oil shock was the failure to draw down stocks near the inception of the crisis. Had a stock drawdown occurred then, perhaps most of the price rise and the economic losses that followed could have been avoided. Governments seem to have argued that no action was needed because a physical shortage did not exist, ignoring the fact that a price spike was forming, which ought to have been the primary signal for a counter-response. Perhaps the IEA countries, acting ultra-cautiously and in our view ill-advisedly, were holding back on stock drawdowns in anticipation of much larger oil losses should war break out. This kind of rationalization, however, would counsel delay in most emergency situations, and indeed it would call into question the usefulness of a stockpile strategy.

An IEA coordinated stock drawdown, used decisively early in the crisis, could be an impressive defensive strategy against

---

21. See Badger, Jr., "International Cooperation during Oil Supply Disruptions," in Horwich and Weimer, *Responding to International Oil Crises*, pp. 1–16.

Table 6-1. *Estimated Strategic Stocks in IEA Countries, 1992, 2000*

Millions of barrels

Stocks	1992	2000
Government-owned		
United States	565	750[a]
Japan	233	315[a]
Germany[b]	205	205
Other	37	40
Usable commercial stocks[c]	210	250
Total	1,250	1,560

Sources: International Energy Agency, "End-December Oil Market Report" (Paris, January 11, 1993), pp. 15–16, 31; and Interagency Working Group, *Strategic Petroleum Reserve: Analysis of Size Options*, DOE/IE-0016 (Department of Energy, February 1990), p. V-4, and chart V-2; and authors' estimates.

a. Assumes announced goals will be fulfilled.

b. Includes stocks owned by a joint government and industry entity, EBV (Erdoel Bevorratungs Verband), which can be controlled by the government. Stocks owned solely by the government amount to 54 million barrels.

c. Excludes the United States, which has no control over commercial stocks, Japan, which is reducing commercial stock requirements as it increases government stocks, and Germany, part of whose commercial stock requirements are held in the EBV, included under government stocks. For all other IEA countries, stocks equal to ninety days of imports are required, but only one-third of these stocks are estimated to be available above operating requirements and thereby usable in the damage response program.

the disruption scenario we have postulated for the year 2000. By then, government stocks are expected to total 1,300 million barrels and usable commercial emergency reserves an additional 250 million barrels (table 6-1). This amount could fully offset the hypothesized loss of 4 mmb/d for a year, or even twice that loss for six months. Strategies for stock releases could vary, as several studies have suggested. One plan might be to undertake to offset three-fourths of the assumed shortfall, which could allow a substantial price rise—say, 25 percent—to restrain demand and provide necessary signals to oil users as safeguards against a disruption lasting longer than expected. Such a strategy could avoid three-fourths of the hypothesized terms-of-trade and disruption costs and yet allow time to decide among diplomatic, economic, and military response options.

What might be done to strengthen the emergency response system in the future? Determining the size of the strategic reserves is of primary concern. In the United States, legislation in 1975 authorized a reserve of a minimum of 500 million and a maximum of 1 billion barrels. The Ford administration chose the minimum, while the Carter administration chose the maximum as a goal, at the same time building toward a reserve of 750 million barrels. The Reagan and Bush administrations supported a goal of 750 million, which is expected to be achieved by the year 2000.[22] Sentiment in the U.S. Congress has supported a target of 1 billion barrels, which probably will keep the question open.

Whether the IEA members that do not have government stocks, especially the larger countries, are willing to begin building them is another open question. Building up government-owned reserves in these countries to an amount equal to, say, thirty days of imports would add about 250 million barrels to high-powered emergency stocks. That would support the principle, still to be agreed on, that government stocks should be drawn first, which would diminish the precautionary and speculative incentives for the private sector to add to its inventories in a crisis. With all governments holding stocks, moreover, a symmetrical response would be facilitated, and contention about "free riders," as well as hesitancy about resolute early action in a crisis, would diminish.

Oil sharing is also in question. The United States would prefer to remove the provision for sharing from the agreement; most other countries favor keeping it. With a strong stock drawdown response as the first line of defense, oil sharing will probably not be needed in most foreseeable circumstances. In any event, this issue does not merit a renegotiation of the IEA treaty, although an informal agreement could usefully resolve differences about how, when, and whether to share oil.

These questions could form the basis for a review of the IEA system and a negotiating agenda on possible changes.

---

22. A U.S. government study, in a careful review of the options, concludes that an increase in the U.S. reserve from 750 million to 1 billion barrels would have a poor cost/benefit ratio. Interagency Working Group, *Strategic Petroleum Reserve: Analysis of Size Options*, DOE/IE-0016 (Washington: Department of Energy, February 1990), pp. 1–7. This study, whose conclusions necessarily depend heavily on the political assumptions chosen, makes no allowance for the effect of changes in U.S. stock policies on those of other IEA members.

IEA-OPEC Consultations

In July 1991, France and Venezuela sponsored a meeting between oil-exporting and importing countries to discuss developments in oil markets. These consultations have been followed by informal meetings between IEA and OPEC experts on technical matters to exchange data and views on medium- and long-term market forecasts, emergency procedures, management of stocks, and related matters.

It is sometimes argued that such discussions could lead to a producer-consumer conference whose purpose would be to ensure stable oil prices in the future. Without belaboring the point, such an effort can be ruled out as impractical and unrealistic. No attempt, past or present, to control or even influence the price of a commodity by international government intervention, has ever been successful.

Nevertheless, much can be said for regular IEA-OPEC discussions. The IEA's defensive strategy to cope with a disruption of supply assumes that some OPEC members having extra capacity will increase their exports. In theory they could choose not to do so or even to reduce exports to offset stock drawdowns by the IEA countries, thus accelerating the surge in oil prices. However, economic, political, and security links between major OPEC exporters and the industrial countries posit a common interest in avoiding destabilization of the oil market. Continued IEA-OPEC exchanges on technical issues can help to deepen this understanding.

Reducing Long-Term Demand

In discussing the outlook for the oil market, we have assumed that present policies affecting oil will remain unchanged. Twenty years is a long time—long enough for environmental considerations, fiscal needs, or technological change to bring about policies that could influence oil markets.

The following frequently cited possibilities could reduce oil demand.

A substantial OECD-wide carbon tax could have important effects. We have argued earlier that in the present state of knowledge and politics such a tax is not to be expected in this decade or the next. However, should future scientific research demonstrate that countering global warming trends is necessary and urgent, OECD countries might reach agreement on a large carbon tax to reduce emissions despite the cost to economic growth. Based on IEA calculations, a tax of $100 a ton on carbon emissions, equivalent to $12 a barrel of oil, could reduce oil consumption in OECD countries in 2005 by about 5 mmb/d.[23]

Among OECD countries, American consumers are the largest and least-taxed users of gasoline. President Bill Clinton's proposed BTU (British thermal units) tax on energy would not change this state of affairs. It would tax crude oil about $3.50 a barrel, equivalent to a tax of slightly over $0.08 a gallon on gasoline. At some point in the future, supporters of a tax might join forces, and arguing for fiscal responsibility, energy security, or protecting the environment might gain approval of a substantial gasoline tax. A tax of $0.75 a gallon, enough to raise the U.S. gasoline tax to barely half the average of the other large OECD countries could reduce U.S. oil consumption in the year 2000 by 1.2 mmb/d, while adding about $75 billion to government revenues.[24]

Technological advance could also make a big difference. The U.S. National Energy Strategy projects possible reductions in oil use of about 3 mmb/d by the year 2010 based on the use of alternative fuels and improvements in propulsion technology. Reductions would be accomplished in part by subsidies, other incentives, and regulations, but this projection assumes there will be substantial increases in the price of oil

---

23. International Energy Agency, *Energy Policies of IEA Countries: 1990 Review* (Paris: OECD/IEA, 1991), pp. 85–87.
24. This estimate of oil savings was obtained from an informal run of a seventy-five-cent gasoline tax on DOE's oil model.

through 2010, which would help to drive these changes. At the oil prices projected in chapter 2, these gains are speculative. However, technological breakthroughs might still occur, most notably in the development of electric automobiles, making these savings more nearly realizable for the United States and resulting in sizable oil conservation in other countries as well.

These possibilities are neither predictable, additive, nor inclusive.[25] To the extent that world demand for oil is reduced by new taxes or technology advances not now taken into account, the net shortfall from a possible interruption in the oil supply would be smaller, as would the potential economic damage it could cause.

25. For example, reforms in energy pricing policies in the former Soviet Union and in the developing countries would bring about sizable reductions in world oil consumption. We have already argued in our projections of oil demand that such reforms will take place at a more rapid rate than commonly believed.

Chapter Seven
Conclusions

WE HAVE ARGUED that the problem of oil security, properly defined, is the risk of a quantum jump in its price. This definition distinguishes oil security from growing dependence on imports or even from a gradual rise in price over a long period, which may be uncomfortable but need not have excessive costs. In turn, the aim of oil security policy is to avoid a sudden, large interruption in the supply of oil, which alone can cause a price shock, and to mitigate the economic and political consequences should it occur.

Our examination of oil markets over the next two decades, including the environmental and political uncertainties that might affect them, is comparatively sanguine in its principal conclusions about oil security. Oil demand should grow moderately, mainly, if not entirely, in response to strong growth in the developing countries. Production capacity should also grow, principally in OPEC countries where the main reserves exist. Oil prices, while continuing to be somewhat volatile in the short run, probably should remain fairly stable over the longer term, staying in a range of $20–$25 a barrel in 1991 dollars through the year 2010.

A principal reason for this promising outlook is the belief that Saudi Arabia and its Persian Gulf oil allies will be able to exert a controlling influence on OPEC policies. These countries have the oil resources to expand production greatly and are strongly interested in avoiding another oil disruption, which would endanger prospects for a growing long-term market for their oil and could incur political and security risks for

86

themselves. Furthermore, experience has shown that the temporary revenue windfalls from oil price jumps disrupt rather than advance orderly economic development.[1]

Nonetheless, another sizable oil disruption has to be rated a possibility, simply because so much of the world's oil needs, a growing proportion in fact, is supplied from an area of chronic political volatility—the Middle East. An interruption could again cause extensive damage to OECD economies and to those of the oil-importing developing countries.

To be sure, several aspects of the prospective oil scene should lessen the economic impact of a future disruption, at least compared with the abysmal experience of the 1970s. Oil use in relation to GDP has declined since then and under foreseeable circumstances should continue to decline through 2010. Also, oil markets have become more efficient. Removal of government price and allocation controls in most OECD countries virtually eliminates the most important distortion in the way markets function in an emergency. The availability of timely, comprehensive data organized by the IEA will make oil markets more transparent in tense circumstances. The existence of a futures market in oil enables buyers and sellers to hedge positions on an increasingly large scale. Finally, the lessons learned from past shocks, including recognition by OECD governments that collaboration and coordination are a necessity, are bound to be a plus. All of these conditions should temper panic reactions when the oil market faces uncertainty about supply.

Even so, a sizable net shortfall in supply from a political upheaval in the Middle East would cause a disproportionately large jump in the price of oil—a response dictated by the inelastic demand for oil in the short run. That conclusion follows from the operation of efficient markets. The only reliable

1. Over the period 1973–85, that is, from the year of the first oil shock to the year just before oil prices collapsed, average growth in GDP for the twelve major oil exporters was approximately one-sixth below the average for all developing countries, calculated from International Monetary Fund, *World Economic Outlook, May 1991* (Washington, 1991), p. 136.

way of reducing the oil price spike and the economic losses following from it is to reduce the size of the net supply shortfall. That would now be possible because the OECD countries have a large volume of unambiguously available emergency oil reserves, which did not exist in the 1970s. The coordinated, timely drawdown of these reserves through the IEA could greatly reduce the size of the shortfall for an extended period, dampen the jump in the price of oil, and moderate the economic losses. It would also allow time to determine whether the causes of the disruption would soon disappear and to consider possible diplomatic, economic, and military options.

Existence of this line of defense is another reason to be comparatively sanguine about the oil outlook. Is this defensive system in need of strengthening? We believe that the size of these OECD emergency oil reserves, specifically those held by governments, should be increased, and the rules for using them improved. Because a new oil shock seems to be a low-probability, even though a high-cost, contingency, governments are reluctant to consider any increase in what amounts to an insurance premium. A tendency to shy away from the difficulties of multilateral decisionmaking, especially the contentious issue of burden-sharing, is also a factor. While not of the highest priority, these matters should not be put aside. Persistence, not urgency, is needed. At modest budgetary cost, the OECD countries could achieve greater economic peace of mind about their oil future and increase the policy flexibility needed to deal with a potentially explosive event. These interests are important for all countries.